Recent Advances in the Application of Mathematical and Computational Models in Biomedical Science and Engineering

Recent Advances in the Application of Mathematical and Computational Models in Biomedical Science and Engineering

Editors

Sundeep Singh
Roderick Melnik
Esther Pueyo

MDPI

Basel • Beijing • Wuhan • Barcelona • Belgrade • Novi Sad • Cluj • Manchester

Editors
Sundeep Singh
Faculty of Sustainable Design Engineering
University of Prince Edward Island
Charlottetown
Canada

Roderick Melnik
Faculty of Science
Wilfrid Laurier University
Waterloo
Canada

Esther Pueyo
Department of Electronic Engineering and Communications
University of Zaragoza
Zaragoza
Spain

Editorial Office
MDPI
St. Alban-Anlage 66
4052 Basel, Switzerland

This is a reprint of articles from the Special Issue published online in the open access journal *Bioengineering* (ISSN 2306-5354) (available at: www.mdpi.com/journal/bioengineering/special_issues/model_Biomedical).

For citation purposes, cite each article independently as indicated on the article page online and as indicated below:

Lastname, A.A.; Lastname, B.B. Article Title. *Journal Name* **Year**, *Volume Number*, Page Range.

ISBN 978-3-0365-9327-2 (Hbk)
ISBN 978-3-0365-9326-5 (PDF)
doi.org/10.3390/books978-3-0365-9326-5

Contents

About the Editors

Sundeep Singh

Dr. Sundeep Singh is working as an assistant professor at the Faculty of Sustainable Design Engineering, University of Prince Edward Island, Canada. His current research expertise is highly multi- and inter-disciplinary, including vast areas in mechanical and biomedical engineering such as thermo-fluids, bio-heat transfer, computational fluid dynamics, computational modeling and simulation of thermal therapies for tumor treatment, and coupled multiscale problems in engineering. He has co-authored over 60 publications in peer-reviewed journals and conferences and serves as a guest editor and reviewer for several reputed journals. He has over 5 years of teaching experience and has received multiple teaching excellence awards.

Roderick Melnik

Prof. Roderick Melnik is a full professor in the Faculty of Science and Tier 1 Canada Research Chair in Mathematical Modelling at Wilfrid Laurier University in Waterloo, Canada. He is the founding director and serves on the board of directors of the MS2Discovery Interdisciplinary Research Institute in Waterloo, Canada. Prior to his appointment in Canada, Prof. Melnik held full professorships in the U.S.A. and Denmark, while in the late 1990s, he held the position of senior mathematician at the Commonwealth Scientific and Industrial Research Organisation (CSIRO) in Sydney, Australia. Prof. Melnik has been a recipient of a number of awards, including prestigious fellowships in Italy, Denmark, England, and Spain. He has published over 900 refereed research contributions as a sole author and with his collaborators from around the world in the fields of applied mathematics, computational sciences, and mathematical modeling in sciences and technology. Prof. Melnik's major results are in the development, analysis, and applications of mathematical models in studying coupled dynamic phenomena, systems, and processes. His research contributions have had a significant impact on a number of areas in nanoscience, bio-nanotechnology, global environmental issues, networks, and more recently, brain research, to name just a few. His interests include human, natural, and man-made systems, along with their interactions. He serves on editorial boards and has been a guest editor for many international journals and book series. He is a member of such professional societies as SIAM, ASME, CAIMS, AustMS, and its ANZIAM Division. Prof. Melnik is head of the M3AI Lab (Mind, Mathematical Models, and AI) and is affiliated with the University of Waterloo and the University of Guelph, as well as with the Guelph-Waterloo Institute of Physics.

Esther Pueyo

Dr. Esther Pueyo currently works as an associate professor at the Department of Electronic Engineering and Communications, University of Zaragoza, Spain. She has participated as a principal investigator and as a collaborator in different international, European, national, and regional research projects and co-authored over 300 publications in peer-reviewed journals and conferences. Her research activity focuses on investigating the mechanisms of arrhythmic risk, either drug-induced or due to pathological effects, with the aim of providing new selective biomarkers that can help to improve the diagnosis of cardiac diseases.

Preface

This reprint is dedicated to the development of novel mathematical models for better understanding complex multidisciplinary interactions and processes at the organ, tissue, and cellular scales for different biomedical science and engineering applications, including diagnostic, therapeutic, imaging, and interventional. We believe that this reprint will provide the readers with ample knowledge and lead to the generation of novel ideas for future advancement in the fascinating field of bioengineering.

Sundeep Singh, Roderick Melnik, and Esther Pueyo
Editors

bioengineering

Article

Robust Control of Repeated Drug Administration with Variable Doses Based on Uncertain Mathematical Model

Zuzana Vitková, Martin Dodek, Eva Miklovičová, Jarmila Pavlovičová, Andrej Babinec and Anton Vitko

Institute of Robotics and Cybernetics, Faculty of Electrical Engineering and Information Technology, Slovak University of Technology in Bratislava, 841 04 Bratislava, Slovakia; zuzana.vitkova@stuba.sk (Z.V.); eva.miklovicova@stuba.sk (E.M.); jarmila.pavlovicova@stuba.sk (J.P.); andrej.babinec@stuba.sk (A.B.); anton.vitko@stuba.sk (A.V.)

* Correspondence: martin.dodek@stuba.sk

Abstract: The aim of this paper was to design a repeated drug administration strategy to reach and maintain the requested drug concentration in the body. Conservative designs require an exact knowledge of pharmacokinetic parameters, which is considered an unrealistic demand. The problem is usually resolved using the trial-and-error open-loop approach; yet, this can be considered insufficient due to the parametric uncertainties as the dosing strategy may induce an undesired behavior of the drug concentrations. Therefore, the presented approach is rather based on the paradigms of system and control theory. An algorithm was designed that computes the required doses to be administered based on the blood samples. Since repeated drug dosing is essentially a discrete time process, the entire design considers the discrete time domain. We have also presented the idea of applying this methodology for the stabilization of an unstable model, for instance, a model of tumor growth. The simulation experiments demonstrated that all variants of the proposed control algorithm can reach and maintain the desired drug concentration robustly, i.e., despite the presence of parametric uncertainties, in a way that is superior to that of the traditional open-loop approach. It was shown that the closed-loop control with the integral controller and stabilizing state feedback is robust against large parametric uncertainties.

Keywords: pharmacokinetics; compartmental models; closed loop control; repeated drug administration; robust control

Citation: Zuzana Vitková, Martin Dodek, Eva Miklovičová, Jarmila Pavlovičová, Andrej Babinec and Anton Vitko Robust Control of Repeated Drug Administration with Variable Doses Based on Uncertain Mathematical Model. *Bioengineering* **2023**, *10*, 921. https://doi.org/10.3390/bioengineering10080921

Academic Editors: Sundeep Singh, Roderick Melnik and Esther Pueyo

Received: 25 June 2023
Revised: 28 July 2023
Accepted: 1 August 2023
Published: 3 August 2023

1. Introduction

The processes running in biological systems, such as in a cell, an organ, or the whole living organism, are very complex, and their dynamics depend on the actual internal states of the system and their interactions; hence, they belong to the category of dynamic systems. In a cell, chains of biochemical reactions typically occur, which can be analyzed and modelled using discrete automata or Petri nets. The main approaches to describe the processes running in the body include the ordinary or partial differential equations, which are abstract mathematical models of the processes. Therefore, the mathematical modeling of the processes in bioscience plays a decisive role in understanding their characteristics and behaviors, which are often hidden. The modelling and control of these processes are based on theoretical results of biocybernetics.

The transport of the drug throughout the body is influenced by a chain of mutually connected and often not fully understood macro and micro processes. Therefore, we usually have significant variability in the choice of modelling structures. A typical uncertainty results from choosing an inadequate structure, due to which some important subprocesses are neglected. This represents the case of the so-called unmodelled dynamics [1].

Another source of uncertainty arises from inaccurate information about the values of model parameters. These include, for instance, the rate constants of the drug absorption from the site of administration to the blood circulation, and the rates of the drug exchange between the body compartments, among others. The mechanism described above is usually referred to as a parametric uncertainty.

The presence of uncertainties leads to their incorporation into the model and in the control design. For that reason, controlled drug delivery, which is required to ensure an optimal therapeutic effect, cannot be resolved solely via experimental sciences. Simply said, the uncertain of in vitro and/or in vivo models naturally calls for an application of cybernetic (system-based) methodologies.

The novelty of the solution presented in this paper is the design of the robust discrete time feedback control system, which can ensure that the drug concentration follows the desired steady state despite the uncertain parameters of the system.

Such applications of system-based approaches to solve problems related to drug kinetics can be found in [2,3]. Additionally, other relevant results have been presented in our recent work [4,5]. In [4], the influence of the surfactant monolaurin of sucrose (MLS) on the rate of absorption was analyzed after an instantaneous per os (peroral) administration to rats of sulfathiazole in the form of suspension. From the in vivo samples of the drug concentrations, the pharmacokinetic model was identified, which was later used to predict the absorption rate constant. The questions related to the role of auxiliary substances in pharmaceutical technology can be found in [6,7].

Based on the available sources and the experiments performed by the authors, the magnitudes of the model uncertainties were assessed and incorporated into the model. Therefore, following the results of [4], in [5], the state-bounding observer was designed for the uncertain model. The observer predicted the guaranteed upper and lower limits of possible drug concentrations in the body compartments, particularly those that are inaccessible for the collection of drug samples. Due to the observer, it was sufficient to take blood samples only from the tissue compartment, and the observer predicted whether the drug concentrations in the other compartments would violate the therapeutic range. The functionality of the observer was demonstrated considering three different models with physiology-based structures. The general theory of state observers can be found in [8–12].

This paper can be considered an extension of the previous results. It attempts to solve the problem of determining the sizes of individual doses that would be sufficiently flexible with respect to the specific requirements of therapy. The requirements also include adaptation of the dosing protocol to the required rate of change in the drug concentration in the body to a desired level and to subsequently maintain it at this level despite uncertain and possibly unstable system dynamics.

There exist various methods to evaluate the influences of the deterministic/random factors and to assess the suitability of the modeling structure for a particular purpose. To mention a few, there are population-based pharmacokinetic models (popPK), linear and nonlinear mixed-effect models (NLME), and physiology-based models (PBM). These also involve methods for determining the suitability of a model, such as Akaike's or Schwarz's information criteria (AIC or SC) [13].

2. Materials and Methods

The in vivo experiment is described in detail in [4,14]. In a nutshell, the experiment was carried out as follows: In the in vivo experiment, 36 rats with a weight of 200 g were considered, which were divided into six groups. Water suspensions containing 5% sulfathiazole were prepared, and at the beginning, a dose of 0.5 mL of suspension per 100 g was administered to all 36 rats. After six hours, using the heparinized injection syringe with cannula, a 2 mL sample of blood was withdrawn from the first group of six animals. The sampling was repeated every hour, but each time for another six animals because the previous group of rats was already dead. Hence, the second group was sampled seven hours after administration, the third group was sampled after eight hours, etc. Then, the mean values of the six samples obtained from each of the six groups were calculated.

In this way, the time dependence of the drug concentrations [mg/mL] versus time [h] was obtained as documented in Table 1. The choice of suspension as a dosage form resulted from the poor solubility of sulfathiazole in water.

Table 1. In vivo concentrations of the drug.

t [h]	0	1	2	3	4	5	6
c [mg/mL]	0	0.0715	0.0855	0.0780	0.0735	0.0490	0.0535

While the value of administered dose is always exactly known, the state variables (concentrations in the compartments) are uncertain to a non-negligible extent due to uncertain parameters. Therefore, the problem calls for a system-based methodology, in particular robust control. To this end, a generalized compartmental model with the parameter Δ determining the uncertainty of the parameters is synthesized and the process is robustly stabilized using the specially designed state feedback.

3. Results

3.1. Continuous Time Model

The method used in this paper is based on the compartmental model shown in Figure 1. The model was parametrically identified using the least squares method.

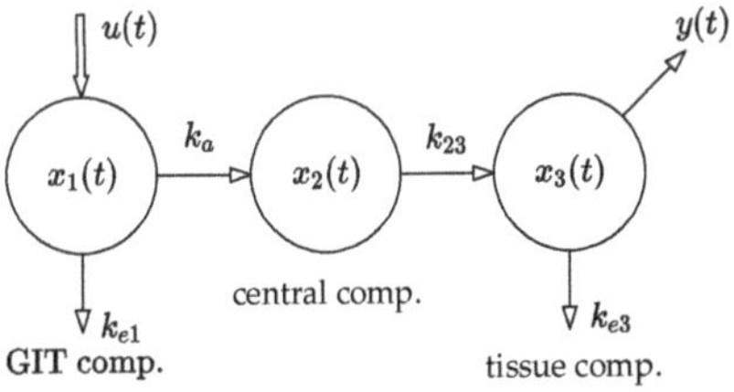

Figure 1. The considered three-compartment pharmacokinetic model showing the drug transport mechanisms, state variables, and model parameters. Variables x_1, x_2, x_3 are the drug concentrations in the gastrointestinal tract (abbr. GIT), central compartment, and tissue compartment, respectively, while u and y are the input and output of the model, respectively.

Let us recall that the general form of the linear system with single input and single output (SISO) is described by the following system of n first-order differential equations:

$$\begin{aligned} \dot{x} &= Ax(t) + bu(t), \\ y(t) &= c^T x(t), \end{aligned} \tag{1}$$

where $x(t) \in \mathbb{R}^{n\times 1}$, $b \in \mathbb{R}^{n\times 1}$, and $c \in \mathbb{R}^{n\times 1}$ are the state, control, and observation vectors, respectively, and $A \in \mathbb{R}^{n\times n}$ is the system matrix, while $y(t) \in \mathbb{R}$ and $u(t) \in \mathbb{R}$ are the output and input of the system, respectively. Generally, if A is a Metzler matrix and vector $b \geq 0$, system (1) belongs to the category of linear positive systems. In addition to that, if A is Metzler and Hurwitz, then system (1) is positive and stable. Specifically, if a positive and stable system respects the rule of mass balance, it is called compartmental [4]. Note that linear and nonlinear compartmental systems are essential for describing the transport of an administered drug throughout the body. In particular, the three-compartment system shown in Figure 1 will be the subject of further analysis in this paper. Its mathematical model takes the following form [5]:

$$\begin{aligned} \begin{pmatrix} \dot{x}_1(t) \\ \dot{x}_2(t) \\ \dot{x}_3(t) \end{pmatrix} &= \begin{pmatrix} -(k_a + k_{e1}) & 0 & 0 \\ k_a & -k_{23} & 0 \\ 0 & k_{23} & -k_{e3} \end{pmatrix} \begin{pmatrix} x_1(t) \\ x_2(t) \\ x_3(t) \end{pmatrix} + \begin{pmatrix} 1 \\ 0 \\ 0 \end{pmatrix} u(t) \\ y(t) &= \begin{pmatrix} 0 & 0 & 1 \end{pmatrix} x(t) \end{aligned} \tag{2}$$

Parameters k_a, k_{e1}, k_{23}, and k_{e3} [h^{-1}] are real positive constants. Parameter k_a quantifies the rate of the drug absorption from the site of its administration. In this case, it is the gastrointestinal tract (abbr. GIT). The values of k_{e1}, k_{e3} are the rates of the drug elimination from the compartments x_1 and x_3, and k_{23} is a measure of the rate of the drug flow from x_2 to x_3.

The relation between the system input $u(t)$—the administered doses—and the system output $y(t)$—the observed concentrations of the drug in the body—is given by the following transfer function [4]:

$$G(s) = \frac{y(s)}{u(s)} = c^T(sI - A)^{-1}b, \tag{3}$$

where "s" indicates that $y(s)$ and $u(s)$ are the Laplace transforms of signals $y(t)$ and $u(t)$ [15,16]. The following nominal parameters were identified in [5]: $k_a = 0.0370\ \mathrm{h}^{-1}$ $k_{e1} = 0.1214\ \mathrm{h}^{-1}$ $k_{23} = 1.2725\ \mathrm{h}^{-1}$ $k_{e3} = 0.2171\ \mathrm{h}^{-1}$.

Remark 1. *After performing the mathematical operations indicated on the right side of (3), we obtain the so-called system transfer function $G(s)$, which defines the relation between the system input and output. It can be expressed in the form of a faction of two polynomial functions (4).*

$$G(s) = \frac{y(s)}{u(s)} = \frac{b_m s^m + b_{m-1}s^{m-1} + \ldots + b_0}{a_n s^n + a_{n-1}s^{n-1} + \ldots + a_0} \tag{4}$$

The polynomial in the denominator is the so-called characteristic polynomial and can be used to determine the system stability. Parameters a_i and b_i include the rate constants k_a, k_{e1}, k_{23}, and k_{e3}.

In contrast to the state (or internal) quantities x_1, x_2, x_3, which are hidden, the values of input $u(t)$ and output $y(t)$ can be directly measured. Therefore, while the parameters a_i, b_i are often easily identifiable, the situation for the constants k_a, k_{e1}, k_{23}, and k_{e3} is more complicated. If we cannot determine them from the known values of a_i, b_i unambiguously, the system is considered "unidentifiable" and the values of the k_a, k_{e1}, k_{23}, k_{e3} must be identified directly from the in vivo samples while considering the system (2). The details can be found in [17,18] and the references therein.

A detailed explanation of the biological meanings of the system parameters, together with their method of identification from the in vivo samples, can be found in [4,5].

Clearly, if the off-diagonal entries of the system matrix A in (2) are non-negative (Metzler matrix), the system (2) is called a positive system. In addition to that, for the nominal parameters, if every column-wise sum is non-positive, the model (1) satisfies the mass balance condition, implying that the system (2) is a nominally stable compartmental system. However, the situation may change when the parameters are subject to uncertainties. Therefore, we check the system stability based on the characteristic polynomial (5).

$$\begin{aligned} \det(sI - A) &= \det\begin{pmatrix} s + k_a + k_{e1} & 0 & 0 \\ k_a & s + k_{23} & 0 \\ 0 & -k_{23} & s + k_{e3} \end{pmatrix} \\ &= (s + k_a + k_{e1})(s + k_{23})(s + k_{e3}) \end{aligned} \tag{5}$$

The direct result of the above polynomial root decomposition shows that all three roots $(s + k_a + k_{e1})$, $(s + k_{23})$, $(s + k_{e3})$ of this characteristic polynomial are negative for any $k_a > 0$, $k_{e1} > 0$, $k_{23} > 0$, $k_{e3} > 0$; hence, the nominal system is stable. This directly implies that the system with uncertain parameters, namely $k_a \pm \Delta k_a$, $k_{e1} \pm \Delta k_{e1}$, $k_{23} \pm \Delta k_{23}$, $k_{e3} \pm \Delta k_{e3}$, will remain stable if $\Delta > -1$. The opposite case would be conditional to the existence of a negative parameter, which is biologically impossible.

3.2. Discrete Time Counterpart of the Continuous Time Model

The discrete time form of the model seems to be more suitable for the problem of finding an appropriate series of repeated drug doses (appropriate protocol). Specifically, for the continuous time system (1), the discrete time counterpart is given by the following difference equation:

$$x(k+1) = Fx(k) + gu(k), \tag{6}$$

where $g \in \mathbb{R}^{n\times 1}$ is the control vector and $F \in \mathbb{R}^{n\times n}$ is the state transition matrix.

Regarding Figure 1, the system output $y(k)$ is given by (7).

$$y(k) = \underbrace{\begin{pmatrix} 0 & 0 & 1 \end{pmatrix}}_{c^T} x \tag{7}$$

Unlike the continuous time model (1), the time t is now represented discretely as a sequence of the time instants $t = kT$, where T is the dosing period. In the in vivo experiment described in [4,14], the dosing period is equal to $T = 6$ h and the same is considered in this paper. The relationship between matrix A of the continuous time system, matrix F of the discrete time system and between vectors b and g is given by (8) [17].

$$\begin{aligned} F &= e^{AT} g = A^{-1}(F - I)b \\ g &= A^{-1}(F - I)b \end{aligned} \tag{8}$$

In an analogy with the continuous time system (1), the discrete time system (6) is positive and stable if the matrix F is non-negative and Schur. Then, considering the dosing interval $T = 6$ h from (8), we can obtain the nominal parameters given in (9).

$$\begin{aligned} F &= \begin{pmatrix} 0.3866 & 0 & 0 \\ 0.0128 & 0.0005 & 0 \\ 0.0718 & 0.3272 & 0.2718 \end{pmatrix} \\ g &= \begin{pmatrix} 3.8726 \\ 0.1025 \\ 0.2704 \end{pmatrix} \end{aligned} \tag{9}$$

The steady-state gain [17], that is the ratio of the steady-state output y_{ss} and the corresponding steady-state input u_{ss} of system (6), equals the following:

$$c^T(I - F)^{-1} g = 1.0759 \tag{10}$$

3.3. Uncertain Discrete Time Model

Thus far, we have supposed that the entries of matrix F and vector g are known exactly, but their actual values can vary within some intervals around the nominal values. Consider the expected maximum acceptable parametric uncertainty $\Delta = \pm\, 0.1$ that is $\pm 10\%$ of their nominal values. Then, the maximal system dynamics will be characterized by matrix $\overline{F}$ and vector $\overline{g}$, as follows:

$$\begin{aligned} \overline{F} &= \begin{pmatrix} 0.4252 & 0 & 0 \\ 0.0141 & 0.0005 & 0 \\ 0.0789 & 0.3599 & 0.2990 \end{pmatrix}, \\ \overline{g} &= \begin{pmatrix} 4.2598 \\ 0.1128 \\ 0.2974 \end{pmatrix}, \end{aligned} \tag{11}$$

and the minimal dynamics will be characterized by matrix $\underline{F}$ and vector $\underline{g}$, as follows:

$$\underline{F} = \begin{pmatrix} 0.3479 & 0 & 0 \\ 0.0115 & 0.0004 & 0 \\ 0.0646 & 0.2944 & 0.2446 \end{pmatrix},$$
$$\underline{g} = \begin{pmatrix} 3.4853 \\ 0.0923 \\ 0.2434 \end{pmatrix}. \tag{12}$$
$$\underline{F} \leq F \leq \overline{F} \text{ and } \underline{g} \leq g \leq \overline{g}$$

Using the Kharitonov theorem [18], it can be easily shown that the uncertain system with the limiting matrices (11) and (12) is stable.

Now, having described the nominal, maximal, and minimal dynamics, we can find the answer to the principal question: Which trajectory $u(k)$, $k = 0, 1, 2, \ldots$ of the repeated drug doses is able to force the output trajectory $y(k)$ to converge to the requested steady-state value $y_{ss} = r$? In general, r is an arbitrary positive real number. Due to the presence of uncertainties, the answer to this question is not straightforward. There are two principal approaches, namely the drug dosing in the open loop and in the closed loop. Both approaches will be briefly explained in the following sections.

3.4. *Drug Dosing in the Open Loop*

In this strategy, the steady-state value of y, i.e., y_{ss} for the unit input $u(k) = 1$, $k = 0, 1, 2 \ldots \infty$ will be determined first. Then, based on y_{ss}, an appropriate dose $u(k)$ will be adjusted such that y_{ss} should be equal to the requested value r. Formally, it can be described as follows:

$$\lim_{k \to \infty} y(k) = y_{ss} = r \tag{13}$$

Hence, the main idea behind the design of the open-loop dosing protocol is as follows: Consider the discrete time model of a general subject shown in Figure 1 and described by the set of difference Equations (6) and (7). Its input $u(k)$ is a sequence of the administrated doses and the output $y(k)$ is the corresponding sequence of the drug concentrations in the body, as shown in Figure 2.

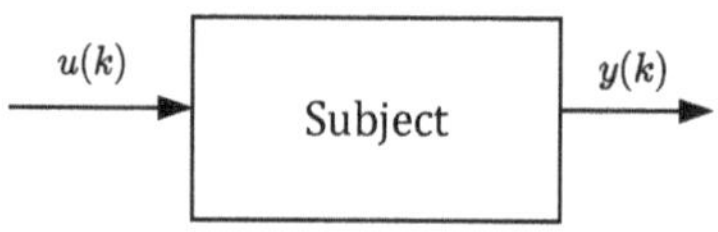

Figure 2. Cybernetic abstraction of a living organism (treated subject) in the case of open-loop dosing approach.

Following the system theory [17], the steady-state value y_{ss} is given by relation (14).

$$r = y_{ss} = \underbrace{\left[c^T (I - F)^{-1} g\right]}_{\text{steady-state gain}} u(k) \tag{14}$$

Note that the expression for the steady-state gain (denoted by the brace in (14)) is nothing more than a scalar multiplier by which the constant unit sequence $u(k) = 1$, $k = 0, 1, 2, 3, \ldots$ of the drug doses should be multiplied to obtain r. Therefore, for a given r and the steady-state gain defined using (14), the following constant drug doses should be repeatedly administered:

$$u(k) = \frac{r}{c^T (I - F)^{-1} g} \tag{15}$$

Note that we will consider the requested steady-state concentration $r = 50$ mg/mL and the sampling period $T = 6$ h. Therefore, the repeated constant doses $u(k)$ should be equal to the following:

$$u(k) = \frac{r}{c^T(I-F)^{-1}g} = \frac{50}{1.0759} = 46.4727 \text{ mg}. \quad (16)$$

For each of the experiments presented further, the area under curve (abbr. AUC) will be determined to quantify the total drug exposure across time considering the experiment length 144 h, nominal models, and using the trapezoidal rule.

The main drawback of this open-loop approach is that the doses are constant in size while presenting no feedback on the current drug concentration in the body. In other words, for a given r, the values of $u(k)$ remain constant regardless of the current drug concentration $y(k)$. Therefore, at least four disadvantages of this approach can be highlighted as follows:

First, the doctor has no means through which they could speed up or slow down the transition process, i.e., the process of gradually approaching the steady-state value $y_{ss} = r = 50$ mg/mL. Second, the transition process of $y(k)$ is unique as it cannot be modified by the constant input sequence $u(k)$. Therefore, there are cases where the transition process can last for an unacceptably long time. The third and most serious drawback is that the behavior of the system induced by the uncertainties cannot be automatically compensated for.

The trajectories of drug concentration $y(k)$ for the calculated constant doses $u(k) = 46.4727$ mg administered every 6 h during the period of 144 h are shown in Figure 3. It can be clearly seen that the nominal trajectory of $y(k)$ (full curve) reaches the requested value $r = 50$ mg exactly and it is maintained within the therapeutic range. Contrary to that, assuming the effect of $\pm 10\%$ uncertainties of the nominal F and g, the steady-state value y_{ss} leaves the therapeutic range (dotted and dashed curves). This means that for model (6), the parametric uncertainties $\pm 10\%$ are too large; hence, the open-loop therapy can significantly jeopardize the patient's health. This is the fourth serious drawback of the open-loop dosing strategy.

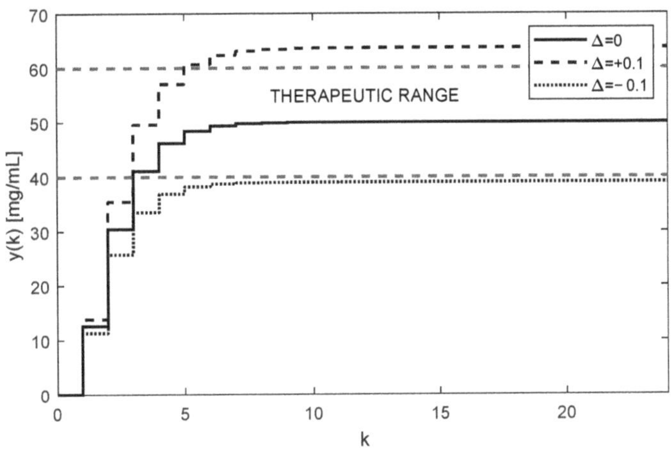

Figure 3. Trajectories of the drug concentrations $y(k)$ in the case of repeated constant dose showing poor robustness of the open-loop approach resulting in either ineffective or toxic treatment (AUC = 6615.7 h × mg/mL).

Since the doctor is not aware of the extent of the parametric uncertainties, the aforementioned drawbacks naturally require a more sophisticated dosing. Much better management of the dosing protocol can be achieved using the closed-loop approach described in the Section 3.5.

3.5. Drug Dosing in the Closed Loop

In this approach, doses $u(k)$ are not constant as they will depend on the current samples of the drug concentrations $y(k)$, for $k = 0, 1, 2, \ldots$. As shown by the closed-loop arrangement, the concentrations $y(k)$ will always converge towards the requested value r regardless of the uncertainties. Due to this property, one can claim that this control strategy is robust with respect to parametric uncertainties [19–21]. The main idea is illustrated in the scheme shown in Figure 4.

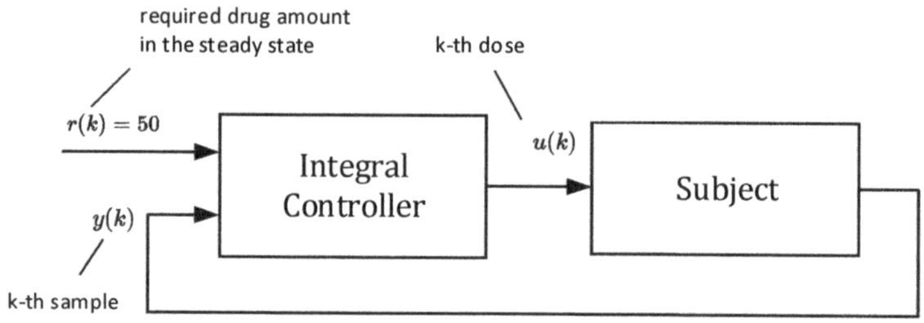

Figure 4. Block diagram of drug dosing in the closed loop that involves the integral controller with feedback on the drug concentration to determine the new dose size.

The block "subject" represents an ill body, which should be treated by repeatedly administrating non-constant doses $u(k)$. To this end, the control algorithm (17) is based on the so-called integral controller. The output of this controller is the sequence of doses $u(k)$, which are to be administered consecutively at the time instants $t = kT$ for $k = 0, 1, 2, \ldots$ [16]. As mentioned above, the role of $u(k)$ is to force the drug concentration $y(k)$ to converge towards the desired steady-state value $y_{ss} = r = 50$ and to maintain it on this level despite the presence of parametric uncertainties.

From the physical and biological nature of the problem, it follows that the quantities $y(-1)$, $u(-1)$, $y(0)$ are equal to zero. The integral controller works according to algorithm (17).

$$u(k) = u(k-1) + k_i[(r - y(k-1)] \text{ for } k = 0,\ 1,2,3, r = 50 \text{ mg} \tag{17}$$

The whole process works as follows: At the beginning, the tunable parameter k_i (the so-called integral gain) [16] is set to some value, for example, $k_i = 0.3$. This value defines not only the rate of increasing the subsequent doses $u(k)$, but also the number of doses required to reach the steady-state concentration $y_{ss} = r$. Then, for $k = 0$, the initial dose is computed as $u(0) = 0 + 0.3(50 - 0) = 15$ mg and the doctor will administrate it. At the end of the first dosing interval ($T = 6$ h), the doctor takes the first sample $y(1)$ and, together with the required steady-state value $y_{ss} = r$, passes this information to the controller (algorithm (17)), which determines the next dose as $u(1) = u(0) + 0.3[50–y(0)] = 15 + 0.3(50 - 4.05) = 15 + 13.78 = 28.78$. The process is repeated for the next indices k. The simulated trajectory of $u(k)$, for $k = 0, 1, \ldots 25$ is shown in Figure 5.

The corresponding trajectory of the drug concentrations $y(k)$ in the site where samples are taken (i.e., from the third compartment) is shown in Figure 6.

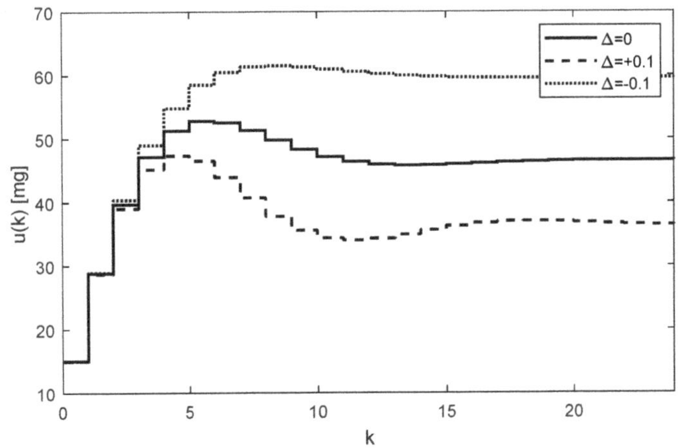

Figure 5. Trajectories of the drug doses $u(k)$ for the integral controller with $k_i = 0.3$ showing a relatively fast response.

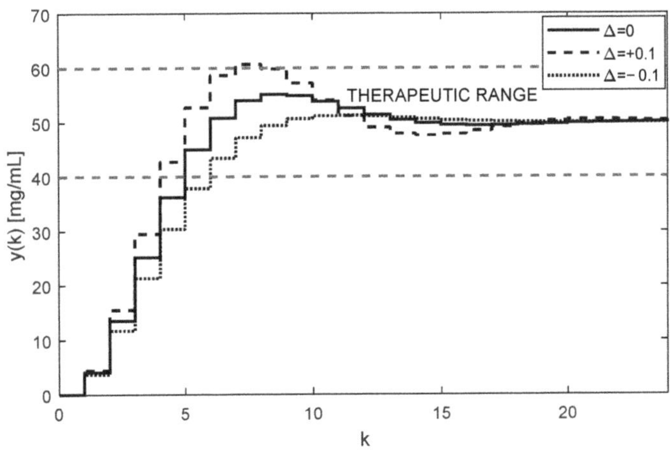

Figure 6. Trajectories of the drug concentration $y(k)$ for the integral controller with $k_i = 0.3$ showing a relatively fast but robust response with slightly periodic behavior and overshoots (AUC = 6419.7 h × mg/mL).

For illustration purposes, the first ten values of the computed doses $u(k)$ and the corresponding trajectory of the drug concentrations $y(k)$ for the nominal case ($\Delta = 0$) are given in Table 2.

Table 2. Drug doses $u(k)$ and trajectory of drug concentrations $y(k)$ for the nominal case.

k	0	1	2	3	4	5	6	7	8	9
$u(k)$	15.00	28.78	39.71	47.14	51.26	52.77	52.52	51.31	49.77	48.30
$y(k)$	0	4.05	13.55	25.24	36.25	44.96	50.82	54.03	55.15	54.88

As Figures 5 and 6 illustrate, both trajectories $u(k)$ and $y(k)$ exhibit some overshoots above their steady-state values. This is caused by choosing an integration gain k_i that is too high. Threfore, it is recommended to decrease its value to $k_i = 0.1$. After carrying this out, more appropriate trajectories are obtained, as shown in Figures 7 and 8.

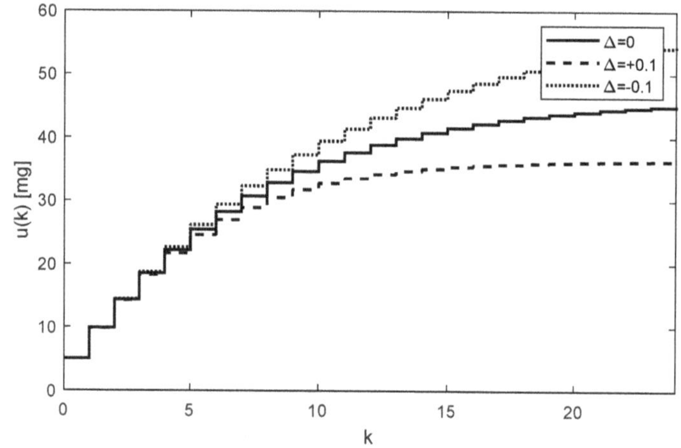

Figure 7. Trajectories of the drug doses $u(k)$ for the integral controller with $k_i = 0.1$ showing a relatively fast response.

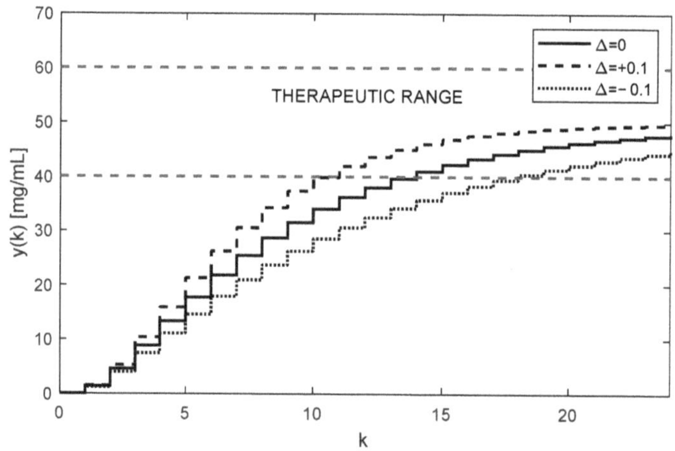

Figure 8. Trajectories of the drug concentrations $y(k)$ for the integral controller with $k_i = 0.1$ showing a relatively slow but robust response with aperiodic behavior and no overshoots (AUC = 4647.5 h × mg/mL).

The nominal trajectory is presented by the full line. The trajectory for positive uncertainties (11) is presented by the dashed curve, and the dot trajectory corresponds to the negative uncertainties (12).

It is even possible to speed up the transition process by intentionally applying a higher loading dose, such as $u(0) = 50$ mg; by carrying this out, we achieved the desired results, as demonstrated in Figure 9.

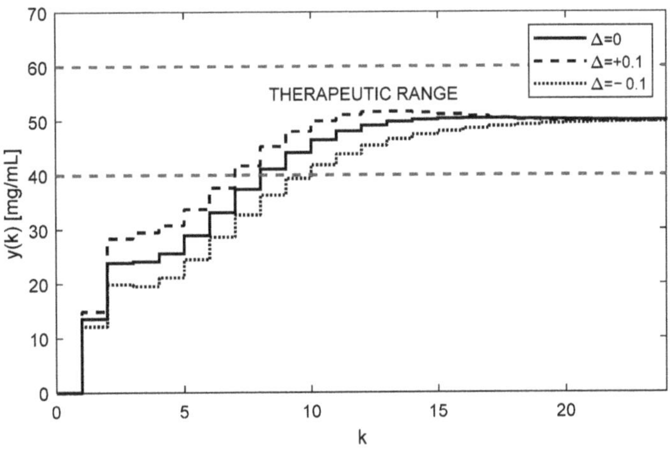

Figure 9. Trajectories of the drug concentrations $y(k)$ for the integral controller with $k_i = 0.1$ and application of a high loading dose $u(0) = 50$ mg (AUC = 5956.0 h × mg/mL).

We can artificially disturb the control configuration and the experiment described above, first, by applying underestimated and, second, by applying overestimated initial doses to make the control algorithm properly correct the dosing. The corresponding response to drug concentration with the first five doses scaled by a factor of 0.1 can be seen in Figure 10, where we can essentially observe a delayed start of the treatment. On the contrary, the response of drug concentration with the first five doses scaled by a factor of 5 can be seen in Figure 11, demonstrating that overestimation can lead to deteriorated performance or even toxic treatment; nevertheless, the control algorithm could ultimately manage this scenario.

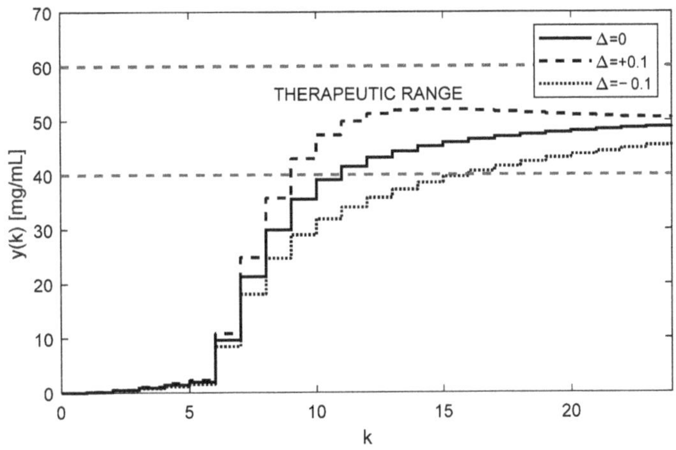

Figure 10. Trajectories of the drug concentrations $y(k)$ for the integral controller with $k_i = 0.1$ and underestimated initial doses (AUC = 4605.0 h × mg/mL).

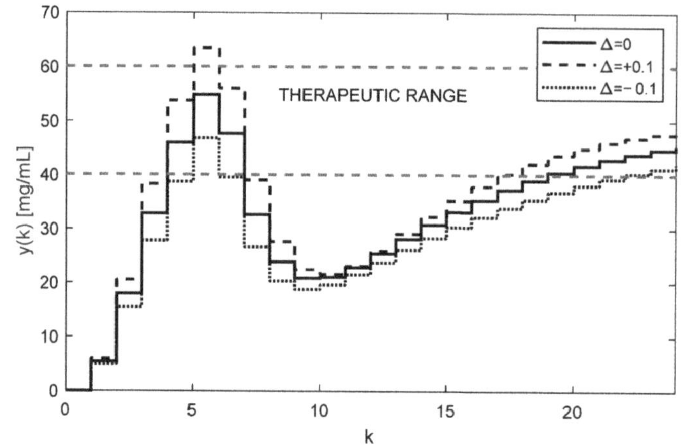

Figure 11. Trajectories of the drug concentrations $y(k)$ for the integral controller with $k_i = 0.1$ and overestimated initial doses (AUC = 4751.2 h × mg/mL).

It is worth mentioning that from the aspect of system theory, the processes of the drug distribution through the body belong to the category of dynamic processes [15]. Therefore, they can be stable or unstable depending on the ways in which the subsystems are mutually connected, as well as the model parameters. The only source of instability of the control structure shown in Figure 4 can arise from an inappropriate (usually too high) integration gain k_i. This was documented by the overshoots in Figures 5 and 6, indicating that the value $k_i = 0.3$ is too high and its further increase may lead to the instability of the closed loop. For that reason, k_i should be chosen to prevent the system from destabilization.

As an illustration, we will check the stability for the considered integral gain $k_i = 0.1$. The transfer function form of the nominal discrete time model (6) and (7) obtains the following:

$$\frac{y(z)}{u(z)} = c^T\left(I - Fz^{-1}\right)^{-1} g z^{-1} \tag{18}$$

Then, we can derive the closed-loop transfer function as follows:

$$\frac{y(z)}{r(z)} = \frac{k_i\, c^T\left(I - Fz^{-1}\right)^{-1} g\, z^{-1}}{\left(1 - z^{-1} + k_i c^T\left(I - Fz^{-1}\right)^{-1} g z^{-2}\right)} \tag{19}$$

For the nominal model with F and g given by (9), we have the following transfer function:

$$\frac{y(z)}{r(z)} = \frac{0.027z^{-1} + 0.027z^{-2} + 0.0003z^{-3}}{1 - 1.6319z^{-1} + 0.7850z^{-2} - 0.1051z^{-3} + 0.0001z^{-4}} \tag{20}$$

The roots of its characteristic polynomial are $\{0.8643,\ 0.5445,\ 0.2226,\ 0.0005\}$.
For $\Delta = 0.1$ with $\overline{F}$ and $\overline{g}$ given by (11), we have the following:

$$\frac{y(z)}{r(z)} = \frac{0.029z^{-1} + 0.025z^{-2} + 0.0004z^{-3}}{1 - 1.6950z^{-1} + 0.8773z^{-2} - 0.1272z^{-3} + 0.0001z^{-4}} \tag{21}$$

The roots of its characteristic polynomial are $\{0.7699,\ 0.6840,\ 0.2406,\ 0.0005\}$.
For $\Delta = -0.1$ with $\underline{F}$ and $\underline{g}$ given by (12), we have the following:

$$\frac{y(z)}{r(z)} = \frac{0.024z^{-1} + 0.0167z^{-2} + 0.0002z^{-3}}{1 - 1.5687z^{-1} + 0.6951z^{-2} - 0.0852z^{-3} + 0.00005z^{-4}} \tag{22}$$

The roots of its characteristic polynomial are $\{0.9037, 0.4606, 0.2039, 0.0004\}$. Since the roots of all characteristic polynomials (20)–(22) lie in the unit circle, the closed-loop control is stable.

The optimal value of k_i should be a trade-off between the speed of increasing the control response of $y(k)$ and the system stability. It is also important to note that, contrary to the open-loop case, the individual doses $u(k)$ are not constant; rather, they gradually increase while their increments decrease as $u(k)$ and $y(k)$ approach their steady-state values. This is evident from Figures 5 and 7. In addition to that, these figures illustrate that at the time instant $t = 24T$, the dose is $u(24) = 46.27$ mg, which is virtually equivalent to the value determined for the open-loop control, i.e., 46.47 mg, from (16).

Another important observation is that Figures 5–8 convincingly demonstrate the robustness of the closed-loop control, meaning that if the parametric uncertainties do not exceed their limits Δ, the system remains stable. Hence, contrary to the open loop, the closed-loop approach with the integral controller ensures that both in the nominal process and in the process disturbed by the parametric uncertainties, the drug concentration $y(k)$ converges to the requested steady-state value $r = 50$ mg/mL, which implies that the control system is robust.

In addition to the advantages of closed-loop dosing mentioned earlier, another practical advantage is the possibility of changing the rate of increasing the trajectory $u(k)$ by changing the controller gain k_i. Clearly, the change in the rate of $u(k)$ is related to the changes in the size of its increments, which, in turn, influence the number of doses needed to reach the requested level r. To see this, compare Figures 5 and 7.

3.6. *Drug Dosing in the Case of an Unstable Subject*

It was shown that the discrete time system (6) is stable for all positive parameters. On the other hand, some special models of bioprocesses, such as physiology-based pharmacokinetic, economic, and ecological models, may not share this feature. Even if their states of equilibria are stable, the region of stability around them can be very small. Therefore, even negligibly small parametric changes or exogenous factors affecting the living subject may render the system unstable. This is manifested by the limitless increase in some characteristic quantities, in particular the system states.

For example, imagine a model of tumor growth, which is typically defined by a set of differential equations [16]. The normally stable tumor-free state equilibrium, i.e., the one computed from the nominal model, can become unstable due to the deviation of the parameters from their nominal values. The instability of the equilibrium will manifest itself through some (or all) variables, e.g., the tumor volume will grow beyond all limits.

However, in this section, we will not deal with any specific model of tumor growth. Instead, we will artificially destabilize the pharmacokinetic model described by (6) and (7). It will be shown that the closed-loop control structure can also be used to control an unstable subject. Consider the feedback scheme shown in Figure 12.

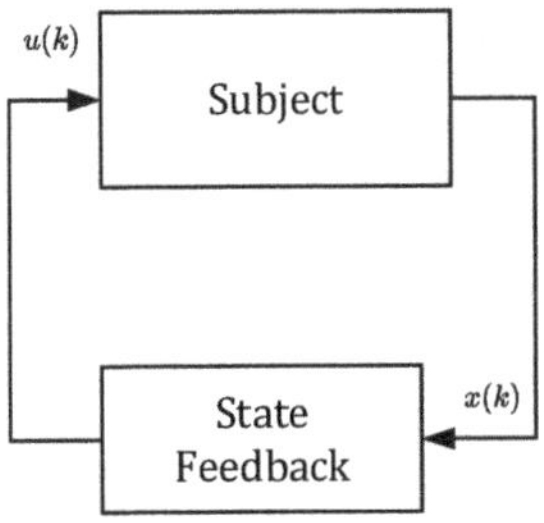

Figure 12. Block diagram of the inner stabilizing closed loop for the drug dosing that involves the state-feedback controller to determine the new stabilizing dose size.

The principle of stabilization via state feedback generally assumes all three components of the state vector $x(k) = [x_1(k),\ x_2(k),\ x_3(k)]$. Therefore, the vector $x(k)$ should contain information on the drug concentrations in all three compartments. As shown in Figure 12, the state vector $x(k)$ is repeatedly sensed and passed to the state feedback where it is converted to the sequence of stabilizing doses $u(k)$.

Obviously, taking drug samples from all three compartments is biologically infeasible; however, their values can be obtained without the need for direct measurement. The states can be generated by the state observer (we have synthesized it in [5], which requires taking only the samples of the output variable $y(k)$). To illustrate this, consider the stable discrete time system described in Equations (6) and (7). To make it artificially unstable, we will intentionally modify the lower uncertainty matrix $\overline{F}$ by adding a destabilizing matrix, as in (23).

$$\begin{aligned}\underline{F} &= \begin{pmatrix} 0.3479 & 0 & 0 \\ 0.0115 & 0.0004 & 0 \\ 0.0646 & 0.2944 & 0.2446 \end{pmatrix} + \begin{pmatrix} 0.7 & 0 & 0 \\ 0 & 0 & 0 \\ 0 & 0 & 0 \end{pmatrix} \\ &= \begin{pmatrix} 1.0479 & 0 & 0 \\ 0.0115 & 0.0004 & 0 \\ 0.0646 & 0.2944 & 0.2446 \end{pmatrix}\end{aligned} \tag{23}$$

The eigenvalues of matrix (23) are as follows:

$$\text{eig}(\underline{F}) = \{0.2446,\ 0.0004,\ 1.0479\} \tag{24}$$

Because the eigenvalue 1.0479 is larger than one, it indicates that the dynamic system (6) with matrix (23) in (9) is unstable [15,17]. The trajectories $u(k)$ obtained using (17) are shown in Figure 13.

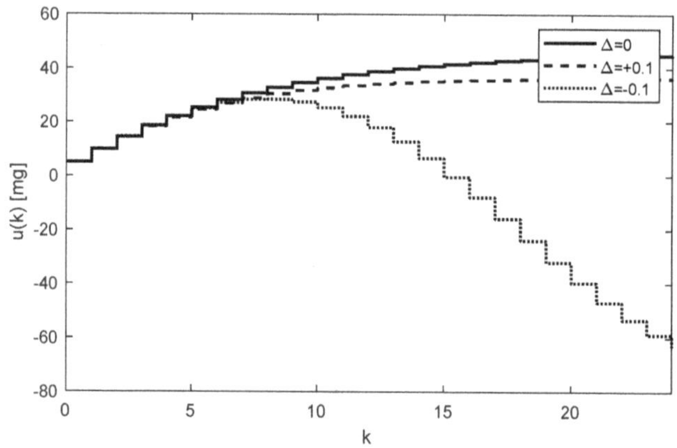

Figure 13. Trajectories of the drug doses $u(k)$ for an unstable subject showing unfeasible negative drug doses to be administered.

It can be observed that for the parametric uncertainty $\Delta = -0.1$, the dotted trajectory $u(k)$ decreases and, at a certain time, becomes negative. Clearly, it can be concluded that the system does not work properly. Therefore, the system should be stabilized via state feedback as proposed. Stabilization was performed by coupling the subject with the stabilizing state feedback $u(k) = Kx(k)$. To determine K, we suppose the interval uncertainties included in (2) and apply Theorem 5 from [22]. To satisfy the conditions of this theorem, we resolved the corresponding problem (25) of linear programming [23].

$$K = K^+ + K^-$$
$$K^+ = \frac{\begin{pmatrix} 1 & 0 & 0 \end{pmatrix} \begin{pmatrix} \sigma_{11}^+ & \sigma_{12}^+ & \sigma_{13}^+ \\ \sigma_{21}^+ & \sigma_{22}^+ & \sigma_{23}^+ \\ \sigma_{31}^+ & \sigma_{32}^+ & \sigma_{33}^+ \end{pmatrix}}{\alpha 1_r{}^T \underline{G}^T v}$$
$$K^- = \frac{\begin{pmatrix} 1 & 0 & 0 \end{pmatrix} \begin{pmatrix} \sigma_{11}^- & \sigma_{12}^- & \sigma_{13}^- \\ \sigma_{21}^- & \sigma_{22}^- & \sigma_{23}^- \\ \sigma_{31}^- & \sigma_{32}^- & \sigma_{33}^- \end{pmatrix}}{1_r{}^T \underline{G}^T v}$$

where

$$\sigma_{ij}^+ < \epsilon^+ \; \forall\, i = \{1,2,3\},\; j = \{1,2,3\}$$
$$\sigma_{ij}^- < \epsilon^- \; \forall\, i = \{1,2,3\},\; j = \{1,2,3\}$$
$$\left(\overline{F}^T - I\right) \begin{pmatrix} v_1 \\ v_2 \\ v_3 \end{pmatrix} + \begin{pmatrix} \epsilon^+ \\ \epsilon^+ \\ \epsilon^+ \end{pmatrix} + \begin{pmatrix} \epsilon^- \\ \epsilon^- \\ \epsilon^- \end{pmatrix} < \begin{pmatrix} 0 \\ 0 \\ 0 \end{pmatrix} \tag{25}$$
$$\alpha \underline{F}\, 1_r{}^T \underline{G}^T v + \underline{G} \begin{pmatrix} \sigma_{11}^+ & \sigma_{12}^+ & \sigma_{13}^+ \\ \sigma_{21}^+ & \sigma_{22}^+ & \sigma_{23}^+ \\ \sigma_{31}^+ & \sigma_{32}^+ & \sigma_{33}^+ \end{pmatrix} + \alpha \overline{G} \begin{pmatrix} \sigma_{11}^- & \sigma_{12}^- & \sigma_{13}^- \\ \sigma_{21}^- & \sigma_{22}^- & \sigma_{23}^- \\ \sigma_{31}^- & \sigma_{32}^- & \sigma_{33}^- \end{pmatrix} \geq \begin{pmatrix} 0 & 0 & 0 \\ 0 & 0 & 0 \\ 0 & 0 & 0 \end{pmatrix}$$
$$\text{chosen } \epsilon^- < 0$$
$$\text{chosen } \epsilon^+ > 0$$
$$\text{chosen } v > 0,\;\; v \in \mathbb{R}^{3\times 1}$$
$$\text{chosen } \alpha > 1$$
$$1_r = \begin{pmatrix} 1 \\ 1 \\ 1 \end{pmatrix}$$
$$\underline{G} \equiv \begin{pmatrix} \underline{g} & \mathbf{0} \end{pmatrix}$$

The obtained feedback gain vector K is as follows:

$$K = \begin{pmatrix} -0.2371 & 0.0754 & 0.0754 \end{pmatrix} \tag{26}$$

The state feedback with the gain K stabilizes the closed-loop system in Figure 12 robustly, that is, under the conditions of uncertain entries of F and g. Actually, by applying the state feedback $u(k) = Kx(k)$, all eigenvalues of the closed-loop system shown in Figure 12 are smaller than one, including the following:

$$\text{eig}(\underline{F} + gK) = \{-0.0054,\ 0.1774,\ 0.2309\} \tag{27}$$

Although all components of vector $x(k)$ are positive, it can be deduced from (26) that one component of the vector K is negative. This means that for some special values of the components of $x(k)$, the product $Kx(k)$ can result in negative doses $u(k)$, which is infeasible. Therefore, the scheme in Figure 12 cannot be used alone. It must be part of the closed loop in Figure 4, as shown in Figure 14. Then, the requested steady-state value r can be flexibly set up to an arbitrary positive value.

Although the original subject (without state feedback) is unstable, the inner closed-loop system shown in Figure 14 is stable. The resulting drug doses $u(k)$ are given by the sum of the controller output $v(k)$ and the feedback signal $fs(k)$ is as follows from (28).

$$u(k) = v(k) + fs(k) = v(k) + K\,x(k) \tag{28}$$

The first ten values of the feedback signal $fs(k)$ and the drug doses $u(k)$ for $T = 6$ h and $k_i = 0.3$ are given in Tables 3 and 4. While the samples of the feedback signals $fs(k)$ are negative, the doses $u(k)$ are positive for any positive value of r. The corresponding trajectories of $y(k) \equiv x_3$ are shown in Figure 15.

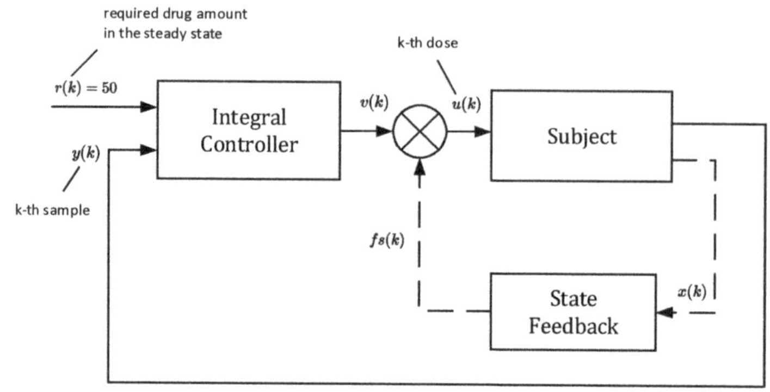

Figure 14. Block diagram of the cascade control loop for the robustly stabilizing drug dosing that involves the state-feedback controller and the integral controller to determine the new dose size.

Table 3. Trajectory of the feedback signal $fs(k)$ for the nominal case.

k	0	1	2	3	4	5	6	7	8	9
$fs(k)$	0	−3.02	−5.25	−7.53	−9.64	−11.65	−13.54	−15.32	−17.00	−18.58

Table 4. Trajectory of the drug dose $u(k)$ for the nominal case.

k	0	1	2	3	4	5	6	7	8	9
$u(k)$	15.00	19.70	25.17	29.21	32.55	35.23	37.40	39.15	40.56	41.70

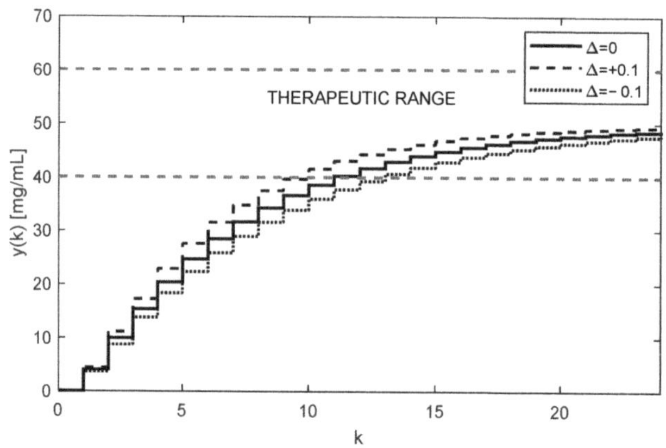

Figure 15. Robustly controlled drug concentrations $y(k)$ of an unstable subject stabilized by the state feedback showing stable behavior despite the unstable nature of the system (AUC = 3737.0 h × mg/mL).

Although the trajectories $y(k)$ displayed in Figures 6 and 15 correspond to the same dosing period and the controller gain ($T = 6$ h, $k_i = 0.3$), they are not quite equivalent. This was expected because, in the previous case, algorithm (16) controlled the stable subject alone, whereas now it controls the subject augmented by the stabilizing state feedback; therefore, it has different dynamics. Regardless of this, by suitably adjusting the integral gain k_i, the shape of the trajectory $u(k)$ can be affected as before.

4. Discussion

The open-loop approach with the dose determined offline can be seen to be theoretically correct and was demonstrated to work properly under the idealized assumption of model parameters that are exactly known. However, it was not possible to prevent the situation where the steady-state concentration of the drug is outside the therapeutic range due to parametric uncertainties. The parametric uncertainties ultimately caused a bias in the system static gain, rendering the determined open-loop dose either ineffective or toxic. Therefore, such a design cannot be considered robust [19,20].

On the other hand, the advantage of the closed-loop approach with the integral controller is that the drug concentrations converged to the requested steady-state value regardless of the actual values of uncertain model parameters. Due to this, for an appropriately chosen controller gain (k_i), the trajectories of the drug concentrations converged to the desired steady-state level (r) aperiodically (i.e., without overshoots). Therefore, any violation of the therapeutic range could be avoided and the therapy was completely safe.

Finally, by adding the state feedback control with an appropriate robust control design, we could achieve the stabilization of a system that became unstable due to uncertainties while ensuring the convergence of the drug concentration to the requested steady-state value. It was shown that after stabilization of the unstable object, it was possible to successfully control the repeated drug administration.

Compared with the fixed dose protocol, the closed-loop delivery scheme is very flexible in affecting the time until the drug concentration reaches the therapeutic range. The performance of the closed loop therapy can be tuned by adjusting the integration gain k_i and the state-feedback gain K in terms of tuning the corresponding parameters of robust control design algorithm (25), which should be chosen in such a way that the monotonic concentration growth is ensured. In this way, the aggressiveness of the therapy can be adjusted to minimize the delays in reaching the therapeutic range by increasing the integration gain. However, a reasonable trade-off between the speed and safety of the therapy must be chosen since more aggressive policies usually lead to dangerous overshoots and, ultimately, to deteriorated robustness. It is important to note that with increasing control aggressiveness, the sensitivity of the control performance with respect to the magnitude of parametric uncertainties also typically increases.

5. Conclusions

This paper considers the pharmacokinetic compartmental model, which the authors designed and parametrically identified from in vivo concentration samples [4,14]. The aim was to design a dosing protocol for repeated administration. It was supposed that the model parameters are uncertain due to numerous reasons, which means that their values vary within a finite interval. The limits of the parametric uncertainties were individually set to 10% of the corresponding nominal values.

Because repeated drug dosing is essentially a discrete time process, it was decided to solve the problem completely in the discrete time domain. To this end, the original continuous time model was first transformed into its discrete time counterpart, and both the open-loop and the closed-loop analyses were performed solely in the discrete time domain. The repeated drug dosing problem was approached in the following three different ways:

First, in the open-loop approach, the system static gain and the doses of constant size, which were repeatedly administered, were determined, forcing the drug concentration to reach the given steady-state level of the drug concentration and to maintain this level for a given time interval.

Second, in the closed-loop approach, the initial oral dose was administered, and after the dosing interval $T = 6$ h, the sample of the drug concentration was taken. Information about its value was passed to the integrating controller, which computed the size of the next dose. This strategy could ensure that drug concentrations converge to the requested steady-state value regardless of actual uncertain model parameters.

In the third approach, the problem of repeated drug dosing for the subject (more precisely, its model), which was not only uncertain but also unstable, was resolved. Although, at first glance, such a situation may seem very rare or even impossible to happen in a living body, it has been shown that this is nothing unnatural. These issues were discussed in Section 3.6.

It can be concluded that the main practical limitation of the proposed strategy is in providing full state feedback; yet, this requirement can be reduced using a state observer [5]. Hence, the required feedback reduces to sensing only the drug concentration in the output compartment, which, today, is feasible due to the extensive development of sensors and wearable electronic devices for biomedical applications.

The results presented in Section 3.6 create the basis for a future analysis of the possible treatment of tumor growth, which is typically an unstable system. This problem is briefly outlined in Appendix A in the form of a short authors' reflection.

Author Contributions: Conceptualization, data acquisition and curation, implementation of in vivo experiments, Z.V. and A.V.; methodology, Z.V.; software, M.D.; validation, Z.V., A.V. and M.D.; formal analysis, E.M.; investigation, A.V. and E.M.; resources, J.P.; writing—original draft preparation, Z.V. and A.V.; writing—review and editing, M.D. and A.B.; visualization, M.D.; supervision, E.M.; funding acquisition, J.P. All authors have read and agreed to the published version of the manuscript.

Funding: The research is supported by the grant VEGA 1/0049/20–Modelling and control of biosystems, granted by the Ministry of education, science, development, and sport of the Slovak republic.

Institutional Review Board Statement: The in vivo experiments on rats were approved by the ethical regulations of the Faculty of pharmacy in Bratislava as a part of the basic research. The certification was sent to the Editorial office.

Informed Consent Statement: Not applicable.

Data Availability Statement: The in vivo data are available from the first author.

Conflicts of Interest: The authors declare no conflict of interest. The funders had no role in the design of the study; in the collection, analyses, or interpretation of data; in the writing of the manuscript, or in the decision to publish the results.

Appendix A

Although the idea shown in Figure 14 is not new from a cybernetic perspective, it can be much more interesting from the aspect of treating diseases via repeated drug dosing. The reader should be aware that from a cybernetic point of view, the subject alone, i.e., without any stabilization feedback, is represented by model (2), which emulates the drug transport through a real living organism.

An important observation is that it is precisely a specific sequence of doses that can stabilize the system. Its specificity means that the sequence is exclusively derived from samples of the drug taken from the patient's body. In other words, the size of every single dose $u(k)$ to be administered is not a choice of the doctor, but is determined by the values of the drug concentrations $x_1(k), x_2(k),\ x_3(k)$ sensed from the different parts of the patient's body, and further modified by both the integral controller and the state feedback. Writing this more formally, it can be summarized as $u(k) = v(k) + fs(k) = v(k) + Kx(k)$, as is shown in Figure 14.

In relation to that, it is worth mentioning that the feedback samples from all three body compartments are not feasible in practice. To resolve this, in [5], we have presented the design of a so-called state observer, which is driven solely by the information on the drug concentration in the body compartment x_3, and the estimates of values $x_1(k)$ and $x_2(k)$ were generated by the observer.

Therefore, if the doctor knows what kind of drug (kind of chemical substance) can treat a particular disease, then a successful treatment will depend on the following two factors:

1. The size of the doses administered at the particular time instants;

2. The order in which the variable doses are administered.

Both parts of that information are provided by the designed control system; however, only the second one plays a decisive role, as is shown in the following:

Let us carry out a short reflection related to the possible ways of treating tumor growth [24,25]. From what has been presented thus far, an idea of how it would be possible to treat tumor growth can be deduced. The issue is that, under normal healthy conditions, the processes (chemical reactions) running in the body are in a dynamic balance. In other words, the processes are in equilibrium. These facts lead the authors to the belief that the conditions under which the tumor starts growing can also be deduced from the behavior of its mathematical model. The start of tumor growth can be a consequence of the existence of either an unstable equilibrium of the mathematical model describing this dynamic, or the existence of a stable equilibrium with a very small region of stability around it.

Under such conditions, it can easily happen that, through the actions of outer influences or flaws occurring inside a living subject, the normally stable equilibrium becomes unstable because the characteristic parameters (e.g., the rates of chemical reactions, and the activity of the enzymes through which the equilibrium is defined, among others) change their values to the extent that the point of equilibria, i.e., its location in the state space, will be outside of the (small) stability region. The same happens in the state space model (2) if the deviations of the parameters cause the equilibrium to move outside its region of stability, as is shown in Section 3.6. At this moment, the system becomes unstable, which is tantamount to the situation when the products of some processes start to increase above all bounds.

It was shown that by applying the control structure shown in Figure 14, it is possible to find a series of repeating doses of various sizes (the dosing protocol) that would be able to stabilize the tumor volume at the given value r. Due to the designed feedback, the tumor volume can be stabilized exclusively through the repeated doses of the specific sizes of an anticancer drug. The current research of the authors is focused on this problem.

References

1. Simkoff, J.M.; Wang, S.; Baldea, M.; Chiang, L.H.; Castillo, I.; Bindlish, R.; Stanley, D.B. Plant–Model Mismatch Estimation from Closed-Loop Data for State-Space Model Predictive Control. *Ind. Eng. Chem. Res.* **2018**, *57*, 3732–3741. [CrossRef]
2. Moradi, H.; Vossoughi, G.; Salarieh, H. Optimal Robust Control of Drug Delivery in Cancer Chemotherapy: A Comparison between Three Control Approaches. *Comput. Methods Programs Biomed.* **2013**, *112*, 69–83. [CrossRef] [PubMed]
3. Ahmed, S.; Özbay, H. Design of a Switched Robust Control Scheme for Drug Delivery in Blood Pressure Regulation. *IFAC-PapersOnLine* **2016**, *49*, 252–257. [CrossRef]
4. Vitková, Z.; Tárník, M.; Miklovičová, E.; Murgaš, J.; Oremusová, J.; Vitko, A. System-Based Approach to Prediction of Surfactants' Influences on Pharmaco-Kinetics and Pharmacodynamics. *Tenside Surfactants Deterg.* **2020**, *57*, 33–39. [CrossRef]
5. Vitková, Z.; Dodek, M.; Pavlovičová, J.; Vitko, A. Using a State-Bounding Observer to Predict the Guaranteed Limits of Drug Amounts in Rats after Oral Administration Based on an Uncertain Pharmacokinetic Model. *Pharmaceutics* **2022**, *14*, 861. [CrossRef] [PubMed]
6. Myers, D. *Surfactant Science and Technology*, 3rd ed.; John Wiley & Sons: Hoboken, NJ, USA, 2005.
7. Ashim Mitra, C.H.L.K.C. *Advanced Drug Delivery*; John Wiley & Sons, Ltd.: Hoboken, NJ, USA, 2016; ISBN 978-1118-022665.
8. Back, J.; Astolfi, A. Positive Linear Observers for Positive Linear Systems: A Sylvester Equation Approach. In Proceedings of the 2006 American Control Conference, Minneapolis, MN, USA, 14–16 June 2006; p. 6.
9. Rami, M.A.; Tadeo, F.; Helmke, U. Positive Observers for Linear Positive Systems, and Their Implications. *Int. J. Control* **2011**, *84*, 716–725. [CrossRef]
10. Härdin, H.M.; van Schuppen, J.H. Observers for Linear Positive Systems. *Linear Algebra Appl.* **2007**, *425*, 571–607. [CrossRef]
11. Gouzé, J.L.; Rapaport, A.; Hadj-Sadok, M.Z. Interval Observers for Uncertain Biological Systems. *Ecol. Modell* **2000**, *133*, 45–56. [CrossRef]
12. Bolajraf, M.; Rami, M.A.; Helmke, U. Robust Positive Interval Observers For Uncertain Positive Systems. *IFAC Proc. Vol.* **2011**, *44*, 14330–14334. [CrossRef]
13. Demidenko, E.; Stukel, T.A. Influence Analysis for Linear Mixed-Effects Models. *Stat. Med.* **2005**, *24*, 893–909. [CrossRef] [PubMed]
14. Vitková, Z.; Tárník, M.; Pavlovičová, J.; Murgaš, J.; Babinec, A.; Vitko, A. In-Vivo Analysis and Model-Based Prediction of Tensides' Influence on Drug Absorption. *Molecules* **2021**, *26*, 5602. [CrossRef]
15. Luenberger, D.G. *Introduction to Dynamic Systems, Theory, Models, and Applications*; John Wiley & Sons: New York, NY, USA, 1979.

16. Norman, S.N. *Control Systems Engineering*; John Wiley & Sons: New York, NY, USA, 2011; ISBN 13 978-0470-54756-4.
17. Williams, R.L.; Lawrence, D.A. *Linear State-Space Control Systems*; John Wiley & Sons: New York, NY, USA, 2007; ISBN 0471735558.
18. Mansour, M.; Kraus, F.; Anderson, B.D.O. Strong Kharitonov Theorem for Discrete Systems. In *Robustness in Identification and Control*; Milanese, M., Tempo, R., Vicino, A., Eds.; Springer: Boston, MA, USA, 1989; pp. 109–124. ISBN 978-1-4615-9552-6.
19. Lu, L.; Yang, R.; Xie, L. Robust H/Sub 2/ and H/Sub /Spl Infin// Control of Discrete-Time Systems with Polytopic Uncertainties via Dynamic Output Feedback. In Proceedings of the American Control Conference, Portland, OR, USA, 8–10 June 2005; Volume 6, pp. 4315–4320.
20. Ackermann, J.; Bartlett, A.; Kaesbauer, D.; Sienel, W.; Steinhauser, R. *Robust Control: Systems with Uncertain Physical Parameters*; Springer: New York, NY, USA, 1993. ISBN 038719843.
21. Lin, F. *Robust Control Design: An Optimal Control Approach*; Wiley: Hoboken, NJ, USA, 2007; ISBN 978-0-470-03191-9.
22. Zhang, J.; Zhao, X.; Zhang, R.; Chen, Y. Improved Controller Design for Uncertain Positive Systems and Its Extension to Uncertain Positive Switched Systems. *Asian J. Control* **2018**, *20*, 159–173. [CrossRef]
23. Bertsimas, D.; Tsitsiklis, J.N. *Introduction to Linear Optimization*; Athena Scientific: Belmont, MA, USA, 1997; ISBN 978-1886529199.
24. Watanabe, Y.; Dahlman, E.L.; Leder, K.Z.; Hui, S.K. A Mathematical Model of Tumor Growth and Its Response to Single Irradiation. *Theor. Biol. Med. Model* **2016**, *13*, 6. [CrossRef] [PubMed]
25. Barbolosi, D.; Iliadis, A. Optimizing Drug Regimens in Cancer Chemotherapy: A Simulation Study Using a PK–PD Model. *Comput. Biol. Med.* **2001**, *31*, 157–172. [CrossRef] [PubMed]

bioengineering MDPI

Article

Evaluation of Mutual Information and Feature Selection for SARS-CoV-2 Respiratory Infection

Sekar Kidambi Raju, Seethalakshmi Ramaswamy, Marwa M. Eid, Sathiamoorthy Gopalan, Faten Khalid Karim and Raja Marappan et al.

[1] School of Computing, SASTRA Deemed University, Thanjavur 613401, India; sekar_kr@cse.sastra.ac.in (S.K.R.); raja_csmath@cse.sastra.edu (R.M.)
[2] Department of Maths, SASHE, SASTRA Deemed University, Thanjavur 613401, India; seethalakshmi@maths.sastra.ac.in (S.R.); sami@maths.sastra.ac.in (S.G.)
[3] Faculty of Artificial Intelligence, Delta University for Science and Technology, Mansoura 11152, Egypt
[4] Department of Computer Sciences, College of Computer and Information Sciences, Princess Nourah bint Abdulrahman University, P.O. Box 84428, Riyadh 11671, Saudi Arabia; dskhafga@pnu.edu.sa
* Correspondence: mmm@ieee.org (M.M.E.); fkdiaaldin@pnu.edu.sa (F.K.K.)

Abstract: This study aims to develop a predictive model for SARS-CoV-2 using machine-learning techniques and to explore various feature selection methods to enhance the accuracy of predictions. A precise forecast of the SARS-CoV-2 respiratory infections spread can help with efficient planning and resource allocation. The proposed model utilizes stochastic regression to capture the virus transmission's stochastic nature, considering data uncertainties. Feature selection techniques are employed to identify the most relevant and informative features contributing to prediction accuracy. Furthermore, the study explores the use of neighbor embedding and Sammon mapping algorithms to visualize high-dimensional SARS-CoV-2 respiratory infection data in a lower-dimensional space, enabling better interpretation and understanding of the underlying patterns. The application of machine-learning techniques for predicting SARS-CoV-2 respiratory infections, the use of statistical measures in healthcare, including confirmed cases, deaths, and recoveries, and an analysis of country-wise dynamics of the pandemic using machine-learning models are used. Our analysis involves the performance of various algorithms, including neural networks (NN), decision trees (DT), random forests (RF), the Adam optimizer (AD), hyperparameters (HP), stochastic regression (SR), neighbor embedding (NE), and Sammon mapping (SM). A pre-processed and feature-extracted SARS-CoV-2 respiratory infection dataset is combined with ADHPSRNESM to form a new orchestration in the proposed model for a perfect prediction to increase the precision of accuracy. The findings of this research can contribute to public health efforts by enabling policymakers and healthcare professionals to make informed decisions based on accurate predictions, ultimately aiding in managing and controlling the SARS-CoV-2 pandemic.

Keywords: SARS-CoV-2 prediction; feature selection; stochastic regression; neighbor embedding; sammon mapping; machine learning

Citation: Sekar Kidambi Raju, Seethalakshmi Ramaswamy, Marwa M. Eid, Sathiamoorthy Gopalan, Faten Khalid Karim and Raja Marappan et al. Evaluation of Mutual Information and Feature Selection for SARS-CoV-2 Respiratory Infection. *Bioengineering* **2023**, *10*, 880. https://doi.org/10.3390/bioengineering10070880

Academic Editor: Mario Petretta

Received: 6 June 2023
Revised: 1 July 2023
Accepted: 13 July 2023
Published: 24 July 2023

1. Introduction

The global health crisis caused by SARS-CoV-2 involving the prediction of respiratory infection and the ultimate causes of the pandemic has highlighted the need for accurate prediction models to aid in effective planning and resource allocation. This study aims to develop a predictive model for SARS-CoV-2 using machine-learning techniques while exploring various feature selection methods to enhance the accuracy of predictions. By incorporating these innovative approaches, valuable insights can be gained into the factors influencing the spread of the virus, ultimately assisting policymakers and healthcare professionals in making informed decisions for managing and controlling the pandemic.

The proposed model employs stochastic regression to capture the stochastic nature of virus transmission and to consider uncertainties in the data. Stochastic regression enables the modeling of complex relationships and provides a comprehensive understanding of the virus's spread dynamics. By leveraging AI-enabled approaches and tools, along with machine-learning-based immune simulation, scientists have created a groundbreaking vaccine resistant to mutations. This vaccine represents a significant breakthrough in the fight against the ongoing COVID-19 pandemic, as it has the potential to provide long-lasting protection against both current and future variants of the virus [1]. This approach acknowledges the inherent uncertainties associated with the pandemic, allowing for more robust and reliable predictions.

Molecular tests, such as reverse transcriptase polymerase chain reaction (RT-PCR), and antigen tests are used to detect current SARS-CoV-2 infection and diagnose COVID-19. These tests have different sensitivity and specificity characteristics, which influence their interpretability. Sensitivity refers to the test's ability to correctly identify positive cases, and specificity indicates its ability to identify negative cases correctly. Understanding the characteristics of diagnostic tests, test timing in relation to symptom onset, and pretest probability of the disease helps in interpreting test results accurately. To analyze the behaviour of DNA probes without regard to mutation, researchers have combined surface-enhanced Raman scattering (SERS) with machine-learning methods. By utilizing SERS, which enhances the Raman scattering signal, and applying machine-learning algorithms, scientists have gained valuable insights into DNA probes' binding characteristics and stability across different SARS-CoV-2 variants. A greater knowledge of the virus and assistance in creating efficient diagnostic tools have been provided by previous authors' illumination of the potential applications of DNA probes in identifying and investigating viral alterations [2]. According to the statistical models present in the research, the regression modeling strategy known as stochastic regression includes randomness or stochasticity in regression analysis. A nonlinear dimensionality reduction method is Sammon mapping, sometimes called Sammon projection. It seeks to maintain the pairwise distances between data points in lower-dimensional space. Divergence is the Bregman measure of the behavioural differences between two probability distributions or vectors.

Additionally, our research explores the use of neighbor embedding and Sammon mapping algorithms to visualize high-dimensional SARS-CoV-2 data in a lower-dimensional space. These visualization techniques allow for better interpretation and understanding of the underlying patterns within the data, facilitating insights into the virus's behavior and transmission dynamics. Developing a point-of-care diagnostic tool that integrates machine-learning algorithms could have significant implications for the timely identification and monitoring of SARS-CoV-2 variants. This could aid in tracking the spread of different variants and implementing appropriate measures to control their transmission [3]. This study considers a range of machine-learning algorithms, including neural networks (NN), decision trees (DT), random forests (RF), and the Adam optimizer (AD), along with hyperparameter optimization techniques (HP). The study aims to predict ACE2 binding and antibody escape, allowing for the identification of potential variants and guiding the development of therapeutic antibody treatments and vaccines for COVID-19 [4]. By evaluating the performance of these algorithms, our research aims to identify the most effective approach for predicting SARS-CoV-2 spread, further enhancing the accuracy of the model. The use of mixed-effects machine learning also helps to predict disease severity in COVID-19 variants of concern, such as Delta and Omicron. These highlight the importance of sequence-level information for the virus in understanding the risks associated with different variants [5].

The active investigation of COVID-19 disease severity prediction is based on the spike protein sequence of SARS-CoV-2. The use of mixed-effects machine-learning techniques shows promise in this area, as it allows for incorporating various factors and variations in data to improve predictive accuracy. These studies highlight the importance of global sequence data, patient outcome information, and early warnings for emerging viral risks [6]. The development of machine-learning approaches is also used to identify key pathogenic regions in coronavirus genomes. Researchers have trained and evaluated millions of models on thousands of genomes, including SARS-CoV-2, MERS-CoV, and others. This study aims to uncover discriminative genomic features that can aid in understanding the pathogenicity of SARS-CoV-2 and other coronaviruses [7]. We obtained complete genome sequences of the Coronaviridae family from the ViPR database and encoded them into binary vector representations. This approach aims to identify genomic characteristics that can differentiate between different coronaviruses, including SARS-CoV-2 [8]. This study highlights the potential of CRISPR-based techniques for rapidly and accurately detecting SARS-CoV-2. We discuss the development of CRISPR-based diagnostic assays that can specifically detect the virus and distinguish it from other coronaviruses [9]. The paper explores early computational detection methods for identifying potential high-risk SARS-CoV-2 variants. It discusses using computational models and algorithms to analyze genetic sequences and predict the impact of specific mutations on virus behavior and transmissibility. By identifying variants of concern early on, this research aims to aid in developing targeted interventions and public health strategies to mitigate the spread of the virus [10].

The suggested model creates a new orchestration from a pre-processed and feature-extracted SARS-CoV-2 dataset and ADHPSRNESM to increase prediction precision and accuracy. This combination of methods and algorithms offers a thorough and reliable framework for predicting the transmission of the infection.

The main objectives in this research work are as follows:

- To focus on developing a predictive model for SARS-CoV-2 using machine-learning techniques, such as neural networks (NN), decision trees (DT), and random forests (RF). By leveraging these algorithms, the researchers aim to capture the virus transmission's stochastic nature and account for data uncertainties.
- To improve the accuracy of predictions, the study employs feature selection techniques. These methods aim to identify the most relevant and informative features contributing to prediction accuracy. By selecting the most important features, the model can focus on the factors that have the greatest impact on the spread of the virus.
- The research explores using neighbor embedding and Sammon mapping algorithms to visualize high-dimensional SARS-CoV-2 data in a lower-dimensional space. This visualization approach enables better interpretation and understanding of the underlying patterns in the data, aiding in identifying important trends and relationships.
- The proposed model combines various techniques, including stochastic regression (SR), neighbor embedding (NE), and Sammon mapping (SM), with a pre-processed and feature-extracted SARS-CoV-2 dataset. The integration of these techniques forms a new orchestration, referred to as ADHPSRNESM, which aims to enhance the precision and accuracy of predictions.

Finally, this research addresses the pressing need for accurate predictive models to better understand and manage the SARS-CoV-2 pandemic. The study aims to improve prediction accuracy and gain valuable insights into the factors influencing the virus's spread by incorporating machine-learning techniques, feature selection methods, and visualization algorithms. The findings of this research have the potential to significantly contribute to public health efforts, enabling policymakers and healthcare professionals to make informed decisions based on reliable predictions, ultimately aiding in the effective management and control of the SARS-CoV-2 pandemic.

Novelty in the Reseach Work

The research work novelty lies in applying feature selection techniques specifically tailored to SARS-CoV-2 data. Traditional feature selection methods often fail to capture the intricate relationships among variables in complex datasets such as that of SARS-CoV-2. Our approach introduces novel feature selection algorithms that consider the unique characteristics and dynamics of the virus, enabling the identification of the most relevant predictors for accurate predictions. We introduce a stochastic regression framework that incorporates probabilistic modeling techniques to account for the inherent randomness and unpredictability of the pandemic. By leveraging stochastic regression, our approach provides more robust and reliable predictions, enabling better decision making in uncertain situations. Neighbor embedding algorithms reduce the dimensionality of the data while preserving the local relationships between instances. Sammon mapping is a nonlinear dimensionality reduction technique that emphasizes preserving the pairwise distances between instances. By leveraging Sammon mapping in our approach, we can effectively capture the intricate relationships and similarities between SARS-CoV-2 cases, improving prediction accuracy and interpretability. Different types of biases, including confounding, selection bias, and measurement error, may affect the validity of observational studies related to SARS-CoV-2. By being aware of these biases, researchers and readers can better interpret and assess the reliability of study findings.

The following is a breakdown of the article. The related work is examined in Section 2, and the suggested model is examined in Section 3. Sections 4 and 5 deal with the experimental setup. In Section 6, the conclusion and future work are presented and followed by refrences.

2. Literature Review and Related Work

The hybrid simulation methods are developed to investigate the effects of mutations on the structural dynamics of the main protease from SARS-CoV-2. The main protease plays a crucial role in viral replication and is a potential target for therapeutic intervention. By combining computational simulations and experimental data, the study provides insights into how mutations affect the protein's stability, function, and interactions with potential inhibitors. This research contributes to understanding the virus's evolution and informs the development of effective antiviral drugs [11]. In this study, the researchers employed supervised learning techniques and perturbation calculations to search for potential inhibitors of the main protease of SARS-CoV-2, which is a crucial target for antiviral drug development. By training a machine-learning model on known inhibitors and non-inhibitors, we can predict the inhibitory activity of new compounds.

Additionally, perturbation calculations were used to assess the compounds' structural stability and binding affinity. This approach holds promise for accelerating the discovery of effective inhibitors against SARS-CoV-2 [12]. This research focuses on whether machines can learn the mutation signatures of SARS-CoV-2 and utilize them to predict the prognosis of infected individuals. The study aims to identify specific viral genotypes associated with varying clinical outcomes by analyzing genomic data from a large cohort of patients. Using machine-learning algorithms, the researchers attempt to create a predictive model that could guide prognosis based on viral genotypes. This work may contribute to personalized medicine approaches for COVID-19 by leveraging viral genetic information [13]. This study investigates the concept of cross-immunity against SARS-CoV-2 variants of concern in naturally infected, critically ill COVID-19 patients. By analyzing patient data, the researchers aim to understand whether prior infection with one virus variant provides any protection or alters the severity of subsequent infections with different variants. The findings from this study could provide valuable insights into immune response dynamics and potential implications for vaccine development and management of COVID-19 patients [14].

This paper proposes a machine-learning-based approach to determine the infection status in recipients of the BBV152 (Covaxin) whole-virion inactivated SARS-CoV-2 vaccine for serological surveys. The authors utilize machine-learning algorithms to analyze serological data and classify individuals as infected based on their antibody response. This approach can aid in evaluating the vaccine's effectiveness and tracking the spread of COVID-19 [15]. In this study, researchers employ machine-learning techniques to predict the docking scores of 3CLpro, a key protein in the SARS-CoV-2 virus. By training a machine-learning model on a dataset of known docking scores, the authors develop a predictive model that can estimate the binding affinity between potential drug compounds and the 3CLpro protein. This work contributes to discovering potential inhibitors for SARS-CoV-2 by prioritizing compounds with favorable docking scores [16]. This research focuses on developing a machine-learning platform that estimates the anti-SARS-CoV-2 activities of compounds. By training a machine-learning model on experimental data regarding the antiviral effects of different compounds, the platform can predict the potential efficacy of new compounds against SARS-CoV-2. This platform could be valuable in the early stages of drug development, helping to identify promising candidates for further investigation [17]. The paper explores the application of machine-learning techniques to predict SARS-CoV-2 infection using blood tests and chest radiographs. It discusses how machine-learning models can analyze patterns and features in these diagnostic tests to accurately predict the presence of the virus. This approach has the potential to enhance early detection and improve patient outcomes [18].

This study uses machine learning and predictive models to present a retrospective analysis of two years of the SARS-CoV-2 pandemic in a single-center setting. The paper investigates various factors related to the spread and impact of the virus, such as demographic information, clinical features, and treatment outcomes. The study aims to provide insights into the long-term trends and patterns associated with the pandemic [19]. The paper proposes a novel approach for determining SARS-CoV-2 epitopes using machine-learning-based in silico methods. Epitopes are specific regions of the virus that can trigger an immune response. By leveraging machine-learning algorithms, this study aims to predict potential epitopes from the viral genome, which can aid in vaccine development and immunotherapy research. The approach offers a promising avenue for identifying crucial targets for immune system recognition [20]. This paper explores machine-learning techniques for screening and diagnosing SARS-CoV-2 (COVID-19) based on clinical analysis parameters. The researchers utilize a dataset containing clinical parameters such as blood test results, vital signs, and patient demographics. By applying machine-learning algorithms, the authors aim to develop a model that can accurately classify individuals as COVID-19-positive or -negative based on these parameters. The study demonstrates the potential of machine learning in assisting with COVID-19 screening by leveraging clinical data [21]. In this paper, the authors propose a diagnostic mask incorporating immunochromatography and machine learning for SARS-CoV-2 detection. The mask is designed to capture and analyze respiratory droplets and provide real-time diagnostic results. By leveraging immunochromatography, the mask detects specific SARS-CoV-2 antigens, and machine-learning algorithms are employed to enhance the accuracy and reliability of the diagnostic process. The integration of these technologies offers a promising approach for rapid and efficient SARS-CoV-2 detection in a wearable form [22].

This paper focuses on the bioactivity classification of SARS-CoV-2 proteinase, an essential enzyme for virus replication, using machine-learning techniques. The researchers employ various machine-learning algorithms to analyze and classify the bioactivity of proteinase inhibitors, which can potentially disrupt the enzymatic activity of the virus. By training the models on a dataset of known inhibitors and their bioactivities, the authors aim to predict the bioactivity of new compounds. This research contributes to developing novel therapeutics by leveraging machine learning in the context of SARS-CoV-2 proteinase inhibition [23]. This paper explores the use of deep-learning techniques to detect the main variants of concern of the SARS-CoV-2 virus. By leveraging the power of deep-learning algorithms, the researchers aim to develop a reliable and efficient method to identify these variants accurately. This could be beneficial for monitoring the spread of different variants and informing public health interventions [24]. In this study, the researchers focus on generating new compounds that specifically target the main protease of the SARS-CoV-2 virus. We address the challenge of imbalanced datasets, which are common in drug discovery research. By employing innovative approaches, such as machine-learning and data augmentation techniques, the researchers aim to overcome this issue and identify potential compounds for further investigation [25]. This paper investigates the application of deep-learning techniques for detecting SARS-CoV-2 using clinical reports. By analyzing a large dataset of clinical reports, the researchers aim to develop a deep-learning model that can accurately identify positive cases of COVID-19. Such a model could potentially assist healthcare professionals in detecting and managing the disease [26]. This research focuses on developing an intelligent system for predicting next-generation sequences of the SARS-CoV-2 virus using deep-learning neural networks. By analyzing existing genomic data, the researchers aim to build a model that is capable of accurately predicting the genetic sequences of future viral strains. This could provide valuable insights into the evolution and behavior of the virus, aiding in the development of targeted interventions and treatments [27–29].

Challenges and Pitfalls in the State of the Art

One limitation of this research is the availability and quality of data. The accuracy and effectiveness of the visualization techniques and machine-learning algorithms we employ heavily depend on the quality and quantity of the SARS-CoV-2 data used. The study's results may be specific to the dataset or population used in the analysis, limiting their applicability to different regions or populations with distinct genetic backgrounds or transmission dynamics. Therefore, caution should be exercised when extrapolating the results to broader contexts. The visualization techniques and machine-learning algorithms employed in the research are based on certain assumptions and simplifications. These assumptions may not fully capture the complexities of the SARS-CoV-2 virus, its mutations, and transmission dynamics. When the model becomes too specialized to the training data, it fails to generalize well to unseen data. To mitigate this limitation, rigorous validation and cross-validation procedures should be applied to ensure the model's robustness and reliability. Reduced dimensionality may limit the ability to fully understand the complex relationships and interactions within the data, potentially hindering a comprehensive interpretation of the virus's behavior and transmission dynamics. Predicting disease severity and understanding of viral risks often involve considering various factors beyond the spike protein sequence, such as co-morbidities, demographics, and environmental factors. Though valuable insights can be gained from these targeted studies, they may not provide a comprehensive understanding of the virus as a whole or address other important aspects, such as the long-term effects, long COVID, or social and behavioral factors influencing transmission. The research might not account for the full spectrum of variants or consider the impact of future mutations that could alter the virus's behavior and transmissibility. Ongoing monitoring and continuous adaptation of the models and algorithms are necessary to keep up with the evolving nature of the virus. Machine-learning models and algorithms have the potential to reinforce biases or yield unintended consequences. It is important to consider the ethical implications of using these techniques in decision making, such as ensuring fairness, avoiding discrimination, and addressing privacy concerns. Though the research aims to evaluate the performance of various machine-learning algorithms, it is crucial to validate their effectiveness in real-world settings. The simulation methods used in the study have their own limitations, including potential inaccuracies and simplifications in the models used.

- Limited and biased datasets: One of the primary challenges is the availability of reliable and representative datasets. Machine-learning models require large and diverse datasets for training, but there may be limitations in terms of data collection and quality.
- There may be inconsistencies or inaccuracies in the reported cases, testing procedures, and other variables. These uncertainties can impact the reliability and robustness of the predictive model.
- Interpretability and visualization: Interpreting and understanding the underlying patterns and relationships in high-dimensional SARS-CoV-2 data can be difficult. Visualization techniques such as neighbor embedding and Sammon mapping can aid

in interpreting and understanding the patterns and relationships within the data, but their effectiveness may vary based on the specific dataset and characteristics.

- Integration of techniques: The proposed model aims to combine various techniques, including stochastic regression (SR), neighbor embedding (NE), and Sammon mapping (SM), to enhance the precision and accuracy of predictions. However, the integration of these techniques introduces additional complexity and potential challenges.

The study's accuracy and generalizability may be affected if the dataset used for training and evaluation is not diverse enough regarding demographics, comorbidities, and other relevant factors. The model's performance may vary when applied to different proteins or protein–ligand systems. The authors should assess the model's performance on diverse datasets to demonstrate its generalizability. The accuracy and reliability of the machine-learning platform may depend on the quality and quantity of experimental data used for model training. Insufficient or biased data may impact the platform's performance. The choice of features used for prediction can impact the model's performance. It is essential to ensure that the selected features are relevant and represent the target variable to avoid bias and improve the model's robustness. The findings of a single-center study may not be generalizable to other healthcare settings, populations, or geographic regions. Replication of the study in multiple centers can strengthen the validity of the results.

3. Methods and Materials

Coronaviruses primarily affect the upper areas of the lungs and respiratory system with different degrees of severity. There is an urgent need for disease prediction at an earlier stage to avoid the causes of death. The IoT has been employed in different application sectors, including healthcare, for SARS-CoV-2 patient monitoring including healthcare. In this case, IoT sensors monitor and gather patient data, which can be accessed anytime and anywhere. Because IoT devices generate vast data, effective prediction is a significant challenge. A proposed model is developed for effective prediction based on the objective. The proposed prediction model for assessing the probability of infection combines classification and feature selection techniques. It uses data mining, stochastic regression, neighbor embedding, Sammon-based map selection, and regression-based classification to predict SARS-CoV-2.

Figure 1 illustrates the basic architecture of this proposed model. The process begins with collecting a significant amount of patient data from IoT devices, which is then stored in a newly created Corona Virus 2019 dataset. The expression "the number of patient records" refers to the count of features in the dataset, which includes both the number of patient records and the count of features represented by these symbols. The collected data undergo several operations using a neural network to facilitate prediction.

The framework of the proposed model is illustrated in Figure 2. The structure consists of several layers, and neurons act as nodes. The nodes in the layers are interconnected to form the entire architecture. The shift-invariant system of the proposed model framework is expressed as:

$$Y(t) = F\,[x(t)] \tag{1}$$

Equation (1) defines a time-dependent output function, a time-dependent input function, and a transfer function used to transfer a transformation or input from one layer to another. The hidden, input, and output layers form machine-learning techniques. The intermediate layers of each feed-forward neural network are called hidden layers and are used to perform certain functions. The output level shows the results of the operation. At first, the features are treated as input, and the activity of neurons in the input layer is presented.

$$x(t) = \sum_{i=1}^{n} f_i(t)\, w_0 + c \tag{2}$$

In Equation (2), $x(t)$ represents the input layer output, $f_i(t)$ shows the features, where 'w_0' denotes the starting weight as "1" and the prime weight as "c" for the bias stored, assigned at the input layer. The first hidden layer receives the input after that.

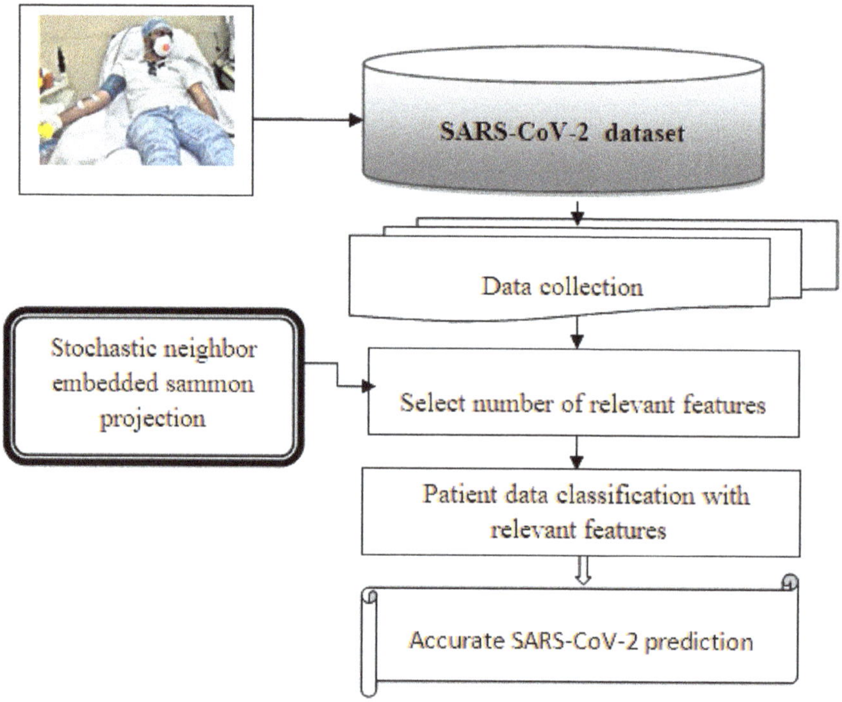

Figure 1. Architecture of the proposed mode.

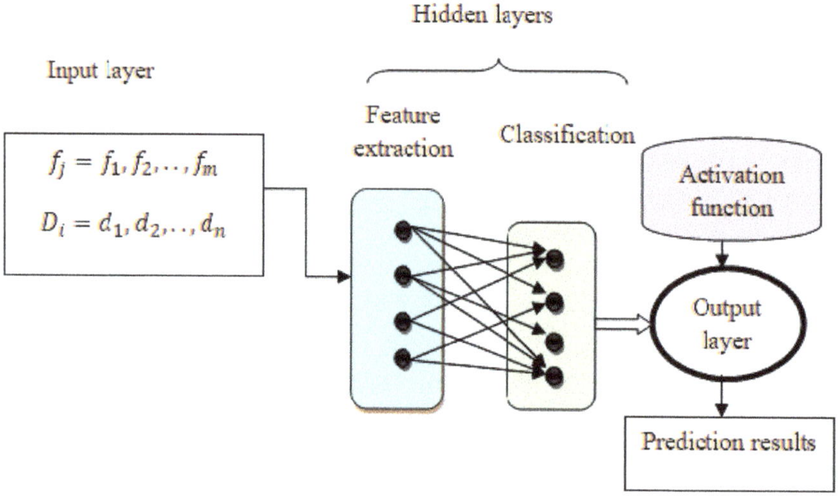

Figure 2. Structure of neural networks with deep connectivity shift-invariant convolution.

3.1. Stochastic Regression

Stochastic regression is a statistical modeling technique that predicts a continuous outcome variable based on one or more predictor variables while incorporating random variation into the model. Unlike traditional regression methods that assume deterministic relationships, stochastic regression acknowledges the presence of inherent uncertainty or random fluctuations in the data. This approach utilizes probabilistic methods, such as maximum likelihood estimation or Bayesian inference, to estimate the parameters of the regression model. By considering the random nature of the data, stochastic regression provides a more realistic and flexible framework for understanding and predicting complex relationships between variables, making it particularly useful in scenarios in which variability and randomness play a significant role.

3.2. Neighbor Embedding

Neighbor embedding is a technique used in machine learning and data analysis to represent high-dimensional data in a lower-dimensional space while preserving the relationships between data points. It aims to map data points onto a lower-dimensional coordinate system, such that neighboring points in the original space remain close in the embedded space. This method is particularly useful for visualization purposes and dimensionality reduction tasks. By organizing and arranging data points based on their proximity, neighbor embedding provides a way to gain insights into the underlying structure of complex datasets.

3.3. Sammon Mapping

Sammon mapping has several advantages and limitations. On the positive side, it can effectively preserve the local structure and relationships between data points. It is particularly useful when dealing with complex, nonlinear data patterns. However, it can be sensitive to outliers and noisy data, which may distort the results. The algorithm's computational complexity can also be high, especially for large datasets. Sammon mapping is a dimensionality reduction technique that aims to preserve pairwise distances when projecting high-dimensional data into a lower-dimensional space. By minimizing the stress function, it seeks to find a representation that faithfully captures the relative distances between data points. Although it has certain limitations, Sammon mapping is a valuable tool for data visualization and exploratory analysis in various domains.

3.4. Bregman Divergence

Bregman divergence is a mathematical concept that measures the dissimilarity between two points in a convex space. It is named after Israeli mathematician L. Bregman, who introduced this measure in the field of convex optimization. Bregman divergence is defined based on a convex function, which serves as a reference for quantifying the difference between the points. It provides a way to compare the distance between two points concerning the chosen convex function, taking into account the geometry of the space. Bregman divergence has various applications in machine learning, data analysis, and information theory, where it is utilized for tasks such as clustering, dimensionality reduction, and optimization. The flexibility of Bregman divergence makes it a valuable tool for capturing and quantifying the dissimilarity between data points in a wide range of applications.

The similar feature selection considers how many features in the provided dataset are spread in the given dimensional space. The distance between the feature and the objective is calculated using the Bregman divergence:

$$\varphi_B = \|f_j - p_o\| \tag{3}$$

Equation (3) represents a Bregman divergence, which is a measure of dissimilarity between probability distributions or vectors. Bregman divergences have various applications in fields such as machine learning, information theory, and data analysis. As a result, the projection output is written as follows:

$$Q \rightarrow \begin{cases} \varphi_B > \delta;\ dissimilar\ features \\ \varphi_B < \delta;\ similar\ features \end{cases} \quad (4)$$

where Q represents a projection function, φ_B represents a Bregman divergence, and δ represents a threshold. The feature that deviates the most from the target is referred to as a different feature. Otherwise, the trait is considered relevant. Similar traits are thus mapped into low-dimensional space in this manner.

In Figure 3, feature selection is a crucial step in machine learning that involves reducing the number of input variables to improve computational efficiency and enhance model performance.

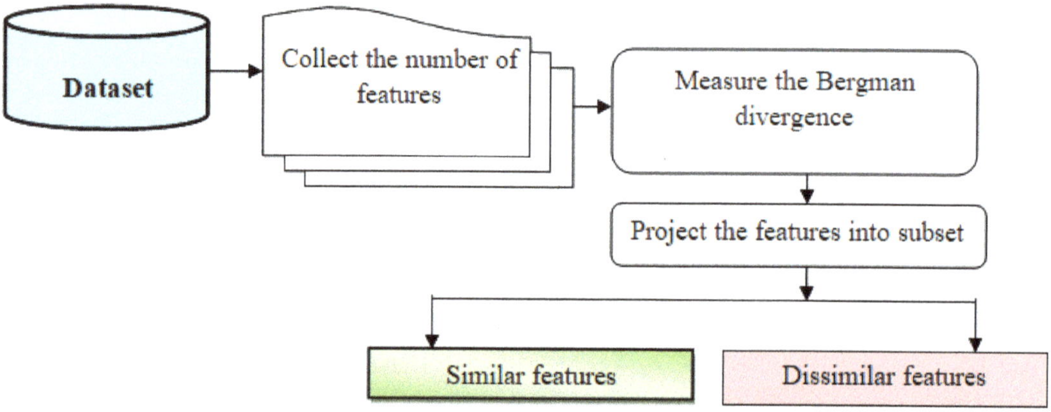

Figure 3. Feature selection procedure block diagram.

3.5. Classification Based on Kriging Regression

Regression analysis, which includes kriging regression, is a collection of statistical methods for estimating the relationships between the variables (i.e., the training and testing features). Based on the association measure between the variables, the patient data are divided into low, medium, and high-risk groups. The regression function looks at the matching feature values, as seen below:

$$R = exp^{\left(\frac{f_{tr} - f_{dt}}{2s^2}\right)} \quad (5)$$

where R represents a regression function, a training feature value, and a disease testing feature value, and s represents a standard deviation. The regression yields values ranging from 0 to 1.

$$Y = \begin{cases} R < 0.5; & low\ risk \\ R = 0.5; & medium\ risk \\ R > 0.5; & high\ risk \end{cases} \quad (6)$$

The patient data are considered high-risk if the regression outcome exceeds the threshold. The patient data are regarded as a medium-risk if the regression value equals the threshold. The patient data are classified as low-risk if the regression exceeds the threshold. Consider the following four training and testing features: age, symptoms such as fever or body temperature, travel history, and chronic condition.

The patient data in Tables 1–3 are categorized as low-, medium-, and high-risk based on the regression analysis with testing and training feature values. Patients at low risk have minimal conditions that can be easily handled. Rapid treatment is required to save the patient's health from the disease if the patient has high-risk conditions.

Table 1. Feature analysis for class 1.

Feature	Age	Symptoms	Travel	Chronic	*Y*	Class 1
Training	>50	37 °C	>28 days	False	<0.5	Low risk
Testing	>50	37 °C	>28 days	False		

Table 2. Feature analysis for class 2.

Feature	Age	Symptoms	Travel	Chronic	*R*	Class 2
Training	>50	37 °C < *T* < 39 °C	14–28 days	True	0.5	Medium risk
Testing	>50	37 °C < *T* < 39 °C	14–28 days	True		

Table 3. Feature analysis for class 3.

Feature	Age	Symptoms	Travel	Chronic	*R*	Class 3
Training	>50	>39 °C	<14 days	True	>0.5	High risk
Testing	>50	>39 °C	<14 days	True		

The hidden layer's outcome is defined as:

$$H(t) = \sum_{i=1}^{n} f_i(t)\ w_0 + [w_{ih} * h_i] \tag{7}$$

where "*H(t)*" stands for the result of the hidden layer, "w_{ih}" stands for the hidden layer's weight, and "h_i" stands for the output of the previously hidden layer that will be sent to the output layer. The outcome of the output layer is:

$$O(t) = w_{ho} * H(t) \tag{8}$$

In Equation (10), the terms "*O(t)*" and "w_{ho}" stand for the output layer result and the weighted average of the hidden and output layers, respectively. To forecast SARS-CoV-2 accurately, the patient data are appropriately divided into several classifications.

The proposed algorithm is given as follows:

Algorithm 1 sketches the SARS-CoV-2 prediction process in detail and more precisely. The DL algorithm comprises many layers that help it learn the provided information. The input features are to be received in its first hidden layer. The Bregman divergence between the features and the goal functions are assessed in that layer. If the divergence exceeds the threshold, the Sammon function separates the characteristics into a comparable and dissimilar subset. Comparable feature subsets are used for classification in the second hidden layer to determine the illness prediction. The patient data are categorized using the Kriging regression method. The regression function contrasts the patient training feature's worth with the sickness testing feature's worth. According to the analysis, a trustworthy sickness has the fastest turnaround time for prognosis.

Algorithm 1: Proposed

Input: Dataset, Number of features $f_j = f_1, f_2, \ldots, f_m$, Number of data $D_i = d_1, d_2, \ldots, d_n$
Output: prediction accuracy increases

1. **Input the number of features** f_j
2. **For every** feature $f_i \in D_i$
3. Measure of the Bregman divergence and between f_i and φ_B
4. **if** ($\varphi_B > \delta$) **then**
5. Project the features as similar
6. **else**
7. Project the features as dissimilar
8. **End if**
9. Select the similar feature subset
10. Remove the dissimilar feature subset
11. **End for**
12. **For** each information in 'd_1' with f_i
13. Perform the regression analysis 'R'
14. **If** ($R < 0.5$) **then**
15. Patient data is classified as 'low risk'
16. **Else If** ($R = 0.5$) **then**
17. Patient data is classified as 'medium risk'
18. **Else If** ($R > 0.5$) **then**
19. Patient data is classified as 'high risk'
20. **End if**
21. Obtain the classification.
22. **End for**
23. **End**

4. Experimental Setup

The proposed model is implemented in Python, and the results are analyzed. For experimentation, Kaggle's Novel Corona Virus 2019 dataset is applied. The daily statistics on coronavirus infections, fatalities, and recoveries are included in the data. There are several CSV files in the collection. Out of all the available files, the open queue file SARS-CoV-2 is applied for testing. This document was downloaded from https://www.kaggle.com/datasets/plameneduardo/sarscov2-ctscan-dataset, accessed on 24 May 2023. The dataset has 306,429 occurrences and eight features such as patient ID, age, gender, city, nation, province, binary code for a chronic condition, symptoms, and travel history. Among these, the key traits are chosen for categorization. To conduct the tests, 1000–10,000 data points are gathered.

5. Results and Discussion

The experimental results of the proposed model are analyzed with the existing models using metrics—prediction accuracy, space and time complexity, and false positive rate [1,2]. The percentage of correct predictions of patient information by the classifier defines the prediction accuracy and is defined as:

$$Acc_{p=}\frac{no\ of\ correct\ predictions}{n} * 100 \tag{9}$$

where n denotes the count of patient information, and the percentage represents the predicted accuracy. The ratio of incorrect prediction to the total records is called the false positive percentage. The number of false positives is determined as follows:

$$R_{FP}= \frac{False\ positive}{True\ negative + \ False\ positive} \tag{10}$$

The count of incorrect predictions signifies the number of patient data that were represented. A percentage (%) is used to represent the false-positive rate. The time it takes the algorithm to forecast the illness is known as the prediction time. The prediction time is, therefore, mathematically expressed as follows:

$$T_p = n * time\ (predicting\ one\ data) \tag{11}$$

where, T_p denotes a prediction time, n denotes the number of data. The prediction time is measured in terms of milliseconds (ms).

$$C_S = n * space\ [\ storing\ one\ data] \tag{12}$$

Table 4 compares the proposed model prediction accuracy to cutting-edge frameworks. As demonstrated in Table 4, compared to the existing models [1,2], the proposed model enhances prediction accuracy for 10,000 distinct instances of unique patient data. The model effectively predicts 890 instances of patient data when the quantity of patient data is set to 1000, whereas [1,2] correctly predict 850 and 820 instances of patient data, respectively. The models in [1,2] have an accuracy rate of 85% and 82%, respectively, compared with 89% in the proposed model. Different numbers of the patient data are input along with a subsequent execution of the numerous runs. The performance of the proposed model is then contrasted with other methods. The average comparison results showed that the proposed model accuracy is 4% greater than that of the current deep long short-term memory (LSTM) ensemble models [1] and 8% higher than that of [2].

Table 4. Comparison with other models.

Patient Data	Prediction Accuracy (%)		
	Proposed Model	**Deep-LSTM**	**MTGP**
1000	89	85	82
2000	91	88	84
3000	90	87	83
4000	91	88	85
5000	90	87	84
6000	92	88	85
7000	91	87	84
8000	90	86	83
9000	91	87	84
10,000	90	86	83

The prediction accuracies of three separate approaches, namely, the proposed model, the Deep-LSTM ensemble model [1], and the multi-task Gaussian process (MTGP) model [2], are represented by three colors: violet, red, and yellow. The proposed model outperforms the other two approaches according to the observed results. The deep connectedness shift-invariant convolutional network was used to examine the training and testing illness features, which resulted in the improvement. The attributes analysis appropriately determines the patient risk prediction level with higher accuracy.

Table 5 shows the experimental findings of the false-positive rate utilizing three methods: the proposed model, the Deep-LSTM ensemble model [1,2], and the MTGP model [3]. The false-positive rate is calculated using a sample size of 10 k patients.

The simulated results show that the proposed model dramatically reduces the FPR compared to the other methods. Consider the 1000 data points for experimentation. The suggested proposed model technique has a false-positive rate of 11, whereas the existing models [1] and [2] have rates of 15% and 18%, respectively. Similarly, for each approach, the following nine outcomes are produced. The proposed model is compared with existing methods using an expected comparison result. The results confirm that the proposed model significantly improves FPR performance by 27% and 41% compared to the other models.

Table 5. Comparison of the FPR.

Patient Data	FPR (%)		
	Proposed Model	Deep-LSTM	MTGP
1000	11	15	18
2000	9	12	16
3000	10	13	17
4000	9	12	15
5000	10	13	16
6000	8	12	15
7000	9	13	16
8000	10	14	17
9000	9	13	16
10,000	10	14	17

The adoption of regression-based classification approaches for predicting SARS-CoV-2 patient data is the source of the improved performance. The proposed model employs Kriging regression to analyze the feature values for the prediction procedure.

The proposed model strategy, on the other hand, is observed to consume less time for SARS-CoV-2 prediction than the other two methods. Stochastic Bregman neighbor-embedded Sammon mapping is used to project the relevant features into the feature subset for illness prediction. The mapping function determines the features that are most relevant to the goal. The categorization uses the selected features and Kriging regression, reducing prediction time.

Table 6 shows the SARS-CoV-2 prediction time utilizing three techniques concerning various patient variables. According to the table values, the proposed model takes less time to forecast the disease level than the ensemble model of Deep-LSTM [1] and the MTGP model [2]. The suggested model technique requires 23 ms of time, while the prediction times for the other two current approaches [1] and [2] are 26 ms and 28 ms, respectively, assuming that there are 1000 patient data. For each strategy, ten outcomes are obtained using various patient data inputs. The prediction time is decreased by 9% and 15%, respectively, as compared to the Deep-LSTM ensemble model [1].

Table 6. Comparison of prediction time.

Patient Size	Prediction Time (in ms)		
	Proposed Model	Deep-LSTM	MTGP
1000	23	26	28
2000	27	32	36
3000	33	36	39
4000	37	40	42
5000	40	45	48
6000	45	48	51
7000	48	53	56
8000	52	56	60
9000	56	59	63
10,000	58	60	65

Table 7 compares the space complexity of three alternative approaches: the proposed model, the Deep-LSTM ensemble model [1], and the MTGP model [2]. The input consists of a variable number of patient data ranging from 1000 to 200, 300, and 1,000,000. Let us say that we have 1000 patient data points to explore. The proposed model takes up the least space for predicting disease, and the Deep-LSTM ensemble [1] and MTGP model [2] take up more. The observed findings demonstrate that the proposed model achieves lower space complexity than existing approaches. Compared to other approaches, the expected result shows that the proposed model reduces space consumption by 9% and 17%.

Table 7. Comparison of space complexity.

Patient Data	Space Complexity (MB)		
	Proposed Model	Deep-LSTM	MTGP
1000	21	24	27
2000	24	26	30
3000	27	30	33
4000	32	36	40
5000	35	40	45
6000	39	43	48
7000	43	46	50
8000	46	50	52
9000	51	54	57
10,000	55	58	61

This is because the SARS-CoV-2 forecast was made before the feature selection procedure was completed. The proposed model uses the relevant features and deletes distinct features. This helps to save space during the SARS-CoV-2 forecast.

In Table 8, the neural network algorithm achieves the highest accuracy, F1-score, recall, and precision among the three algorithms. It attains an accuracy of 98.78%, an F1-score of 99.10%, and a recall and precision of 98.92%. The neural network also exhibits a relatively low error rate of 0.0122. Additionally, the neural network algorithm consumes 3.278726674 units of time for execution. However, it is worth noting that the loss function plot for the neural network model appears unusual, with the validation dataset's accuracy remaining higher than the training dataset and the validation loss lower than the training loss, which is contrary to expectations.

Table 8. Machine-learning techniques with core analysis.

Algorithm	Accuracy	F1-Score	Recall	Precision	Time Consumption	Loss Function	Error Rate
Neural network	97.78%	98.80%	98.32%	98.28%	4.688728674	46,934.55218	0.02222
Decision tree	73.59%	73.44%	73.59%	73.60%	6.488522147	74,401.98414	0.2641
Random forest	72.67%	72.15%	72.66%	72.68%	3.781597534	57,045.66329	0.2733

In Table 9, the decision tree algorithm achieves a moderate accuracy of 74.59%, an F1-score of 74.56%, a recall of 74.64%, and a precision of 74.42%. The decision tree algorithm has a higher error rate of 0.2541 and takes 5.688512147 units of time for execution. The random forest algorithm attains an accuracy of 72.99%, an F1-score of 73.35%, recall of 73.44%, and precision of 73.36%. It has a similar error rate as the decision tree algorithm (0.2701) but a faster execution time of 2.582577534 units. These results indicate that the neural network algorithm outperforms the decision tree and random forest algorithms in terms of accuracy, F1-score, and precision, while also achieving a low error rate.

Table 9. Dataset with Adam optimizer.

Algorithm	Accuracy	F1-Score	Recall	Precision	Time Consumption	Loss Function	Error Rate
Neural network	98.78%	99.10%	98.92%	98.92%	3.278726674	34,934.45318	0.0122
Decision tree	74.59%	74.56%	74.64%	74.42%	5.688512147	64,321.88314	0.2541
Random forest	72.99%	73.35%	73.44%	73.36%	2.582577534	48,035.56229	0.2701
Proposed model	99.03%	99.32%	98.98%	99.06%	2.12368745	33,876.36407	0.0118

However, further investigation is necessary to understand the anomalous loss function plot observed for the neural network model. The proposed model demonstrates exceptional performance with high accuracy, precision, and recall, outperforming all the other models mentioned. The time consumption for this model was approximately 2.12 s. Finally, based on the provided information, the proposed model outperforms the decision tree and random forest models regarding performance metrics such as accuracy, precision, and recall. However, it is important to note that the absence of the specific loss function used for these models limits the analysis of their training process and further comparison.

In Table 10, the neural network algorithm exhibits exceptional performance across all evaluated metrics compared to the decision tree and random forest algorithms. The neural network algorithm achieves an accuracy of 99.22%, an F1-score of 99.31%, a recall of 99.33%, and a precision of 99.23% [1]. These results indicate a high accuracy, precision, recall, and F1-score, showcasing the algorithm's effectiveness in accurately classifying data. On the other hand, the decision tree algorithm demonstrates moderately good performance with an accuracy of 75.03%, an F1-score of 74.77%, a recall of 75.05%, and a precision of 74.73% [2]. Although these metrics suggest reasonable performance, they fall short compared to the neural network algorithm.

Table 10. Dataset with Adam optimizer and hyperparameters.

Algorithm	Accuracy	F1-Score	Recall	Precision	Time Consumption	Loss Function	Error Rate
Neural network	99.22%	99.31%	99.33%	99.23%	2.818727	34,933.2532	0.78
Decision tree	75.03%	74.77%	75.05%	74.73%	5.228512	64,320.6831	0.2497
Random forest	73.43%	73.56%	73.85%	73.67%	2.122578	48,034.3623	0.2701
Proposed model	99.41%	99.46%	99.45%	99.37%	2.032367	33,842.2328	0.2391

Similarly, the random forest algorithm also exhibits decent performance. Still, it is slightly lower than the decision tree algorithm, with an accuracy of 73.43%, an F1-score of 73.56%, a recall of 73.85%, and a precision of 73.67% [2]. Overall, the results highlight the neural network algorithm as the most effective and reliable choice among the three algorithms. However, it is essential to conduct a comprehensive analysis considering additional metrics such as time consumption, loss function, and error rate to understand the algorithms' performance fully. Based on these comparisons, the proposed model demonstrates superior performance in accuracy, F1-score, recall, precision, time consumption, loss function, and error rate. However, it is important to note that the specific context and requirements of the problem should also be considered when selecting the most suitable model for a given task.

In Table 11, showing the proposed model, as well as the neural network, decision tree, and random forest algorithms, it is evident that each model performs differently across the various evaluation measures. The proposed model's mean absolute error (MAE) is 0.0163923 [3]. In comparison, the neural network achieves a slightly higher MAE of 0.0236756 [2]. The decision tree model yields a MAE of 0.0174911, and the random forest model obtains a similar MAE of 0.0226886 [1]. Moving on to the mean squared error (MSE), the proposed model achieves the lowest value at 0.00131245, followed by the neural network with 0.00149826. The decision tree model has a MSE of 0.0026259, and the random forest model performs slightly better with a MSE of 0.0017358. The root mean squared error (RMSE) for the proposed model is 0.03499126, and the neural network and decision tree models have RMSE values of 0.03649006 and 0.0503219, respectively. The random forest model performs the best in terms of RMSE, achieving a value of 0.0413318.

When considering the coefficient of determination (R-squared), which measures the proportion of the variance in the dependent variable that the independent variables can explain, the random forest model outperforms the other models with an R-squared value of 0.9886125. The proposed model and neural network follow closely with R-squared values of 0.97114391 and 0.97315592, respectively. The decision tree model achieves an R-squared value of 0.9786136.

Table 11. Dataset with Adam optimizer, hyperparameters with stochastic regression.

Metrics	Proposed Model	Neural Network	Decision Tree	Random Forest
MAE	0.0163923	0.0236756	0.0174911	0.0226886
MSE	0.00131245	0.00149826	0.0026259	0.0017358
RME	0.03499126	0.03649006	0.0503219	0.0413318
R-Squared	0.97114391	0.97315592	0.9786136	0.9886125
RAE	0.0868276	0.1258386	0.0978889	0.0977898
RRSE	0.1613141	0.1769975	0.1713163	0.1624251

The relative absolute error (RAE), which assesses the average difference between predicted and actual values relative to the mean of the actual values, is the lowest for the proposed model at 0.0868276. The neural network model has a higher RAE of 0.1258386, and the decision tree and random forest models yield RAE values of 0.0978889 and 0.0977898, respectively. Lastly, the relative root squared error (RRSE), which measures the average difference between predicted and actual values relative to the range of the actual values, is the lowest for the proposed model at 0.1613141. The neural network, decision tree, and random forest models achieve RRSE values of 0.1769975, 0.1713163, and 0.1624251, respectively.

Finally, based on the provided metrics, the proposed model showcases the best performance in terms of MAE and MSE. However, the random forest model outperforms the other models regarding R-squared, indicating a higher proportion of explained variance. Additionally, the proposed model demonstrates lower RAE and RRSE values, suggesting better accuracy than the other models.

In Table 12, showing the neural network, decision tree, random forest, and proposed models, we can observe variations in the mortality rate, vaccination rate, and other relevant features. The neural network model demonstrates a mortality rate of 0.05 and a vaccination rate of 0.02, and other relevant features are at 0.7. The decision tree model, on the other hand, shows a lower mortality rate of 0.02 and a vaccination rate of 0.01, with other relevant features at 0.85. Comparatively, the random forest model has a higher mortality rate of 0.1 and a vaccination rate of 0.03, and other relevant features are at 0.5. The proposed model falls in between, with a mortality rate of 0.056, a vaccination rate of 0.04, and other applicable features at 0.78.

Table 12. Dataset with Adam optimizer, hyperparameters with stochastic regression with Sammon mapping.

Algorithm	Mortality Rate in %	Vaccination Rate in %	Other Relevant Features in %
Neural Network	0.05	0.02	0.7
Decision Tree	0.02	0.01	0.85
Random Forest	0.1	0.03	0.5
Proposed Model	0.056	0.04	0.78

From these results, it can be inferred that the decision tree model has the lowest mortality rate of 0.02, followed by the proposed model with a mortality rate of 0.056. The neural network and random forest models demonstrate higher mortality rates of 0.05 and 0.1, respectively. Regarding vaccination rates, the decision tree model has the lowest value at 0.01, and the random forest model has the highest value at 0.03. The proposed model falls in between, with a vaccination rate of 0.04. When considering other relevant features, the decision tree model has the highest value at 0.85, followed by the proposed model with 0.78. The neural network model has other relevant features at 0.7, and the random forest model has the lowest value at 0.5.

At last, based on the provided data, the decision tree model showcases the lowest mortality rate and the highest vaccination rate among the algorithms. However, the proposed model performs reasonably well, with a mortality rate higher than the decision tree model but lower than the neural network and random forest models. Additionally, the proposed model demonstrates a moderate vaccination rate and comparatively high values for other relevant features.

A widely used approach to evaluating the performance of a machine-learning model is a confusion matrix. This matrix provides a tabulated representation of the predicted results versus the actual outcomes, which allows for a comprehensive examination of the model's effectiveness and limitations. By analyzing the matrix, one can obtain a more detailed understanding of the model's strengths and weaknesses, which can help in making informed decisions about improving its performance in Table 13. A statistical analysis of distributions of data involves using various statistical methods to describe and summarize the characteristics of a dataset. The main objective of this analysis is to gain insight into the underlying patterns and trends within the data. These measures provide information about the typical or average values of the dataset in Table 13.

Table 13. Distributions of the data.

Statistical Measure	Confirmed	Deaths	Recovered
count	3.06×10^3	306,429	3.06×10^5
mean	8.567091×10^4	2036.403268	5.04×10^4
std	2.78×10^5	6410.938048	2.02×10^5
min	-3.03×10^5	−178.000000	-8.54×10^5
25%	1.04×10^3	13	1.10×10^1
50%	1.037500×10^4	192	1.75×10^3
75%	5.08×10^4	1322	2.03×10^4
max	5.86×10^6	112,385	6.40×10^6

Among the countries most affected by the SARS-CoV-2 pandemic, the top 10 confirmed cases include the United States, India, Brazil, Russia, the United Kingdom, France, Turkey, Italy, Spain, and Germany. Though some nations have seen a decline in new infections and deaths, others struggle with high transmission rates and overwhelmed healthcare systems. In terms of recovery rates, many countries have shown significant improvements thanks to the availability of vaccines and better treatment options. However, disparities still exist based on factors such as age, underlying health conditions, and access to healthcare Figure 4.

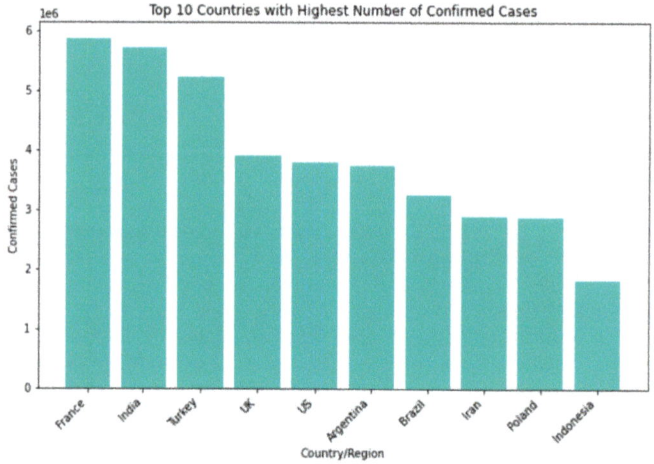

Figure 4. Top 10 countries with highest number of confirmed cases.

A confusion matrix is a table that is used to evaluate the performance of a machine-learning model by comparing predicted and actual outcomes. It provides a useful way to visualize the true positives, true negatives, false positives, and false negatives generated by a model. Neural networks are a machine-learning model inspired by the structure of the human brain. They consist of layers of interconnected nodes that perform mathematical operations on input data, gradually learning to recognize patterns and make predictions. Decision trees are another type of machine-learning model that work by recursively splitting data into smaller subsets based on the most informative features. They create a tree-like structure that can be easily visualized, making them useful for explaining how a model makes decisions. Random forests are an ensemble method that combine multiple decision trees to improve predictive accuracy. Each tree in the forest is built on a random subset of the training data and a random subset of the available features. This helps to reduce overfitting and increase the generalization ability of the model. In Figure 5, death and recoveries over time are exhibited using the metrics of std, mean and count. In Figures 6–8, shows the error rate is a metric used to measure the performance of a machine learning model. It quantifies the number of incorrect predictions made by the model. Loss functions play a vital role in machine learning models as they quantify the discrepancy between predicted outputs and the actual values. Computational time is an important consideration when working with machine learning models. The time required for training and inference can vary depending on the complexity of the model, the size of the dataset, and the available computational resources.

To evaluate the performance of a neural network, a decision tree and random forest use a ROC curve; the network is trained on a labeled dataset and then tested on a separate set of labeled instances. The predicted labels and actual labels are compared to calculate the TPR and FPR at different threshold values, and these values are plotted on the ROC curve in Figures 9–11. Finally, the ROC curve is a useful tool for evaluating the performance of binary classifiers such as neural networks, decision trees, and random forests. Plotting the TPR and FPR at different threshold values shows how the classifier's performance changes as the threshold is adjusted, allowing for easy comparison between different classifiers.

The vast and complex world of machine learning requires a deep understanding of various models, performance metrics, and optimization techniques. To improve the performance of machine-learning models, it is essential to experiment with different hyperparameters, regression techniques, and embedding methods to fine-tune them. By carefully optimizing these factors, our machine-learning models can achieve greater accuracy and efficiency.

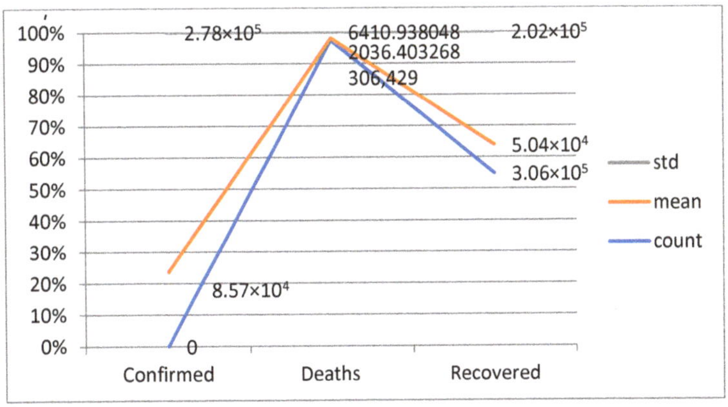

Figure 5. Death and recoveries over time.

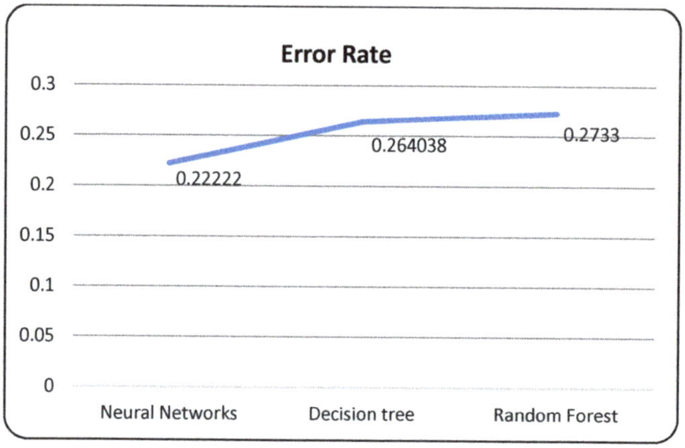

Figure 6. Models with error rate.

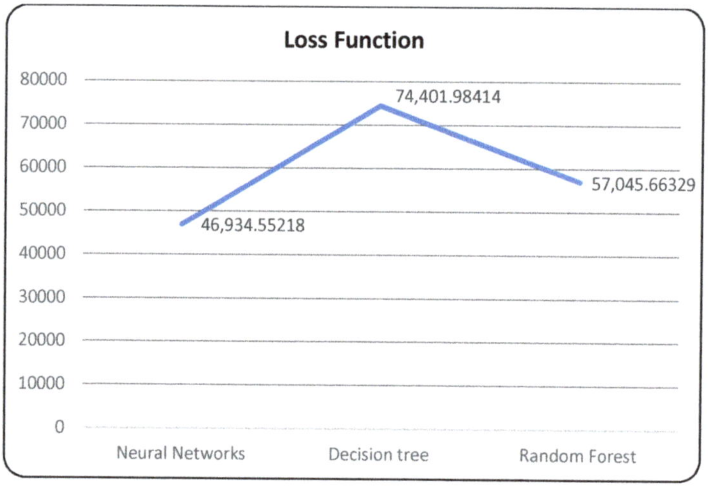

Figure 7. Models with loss function.

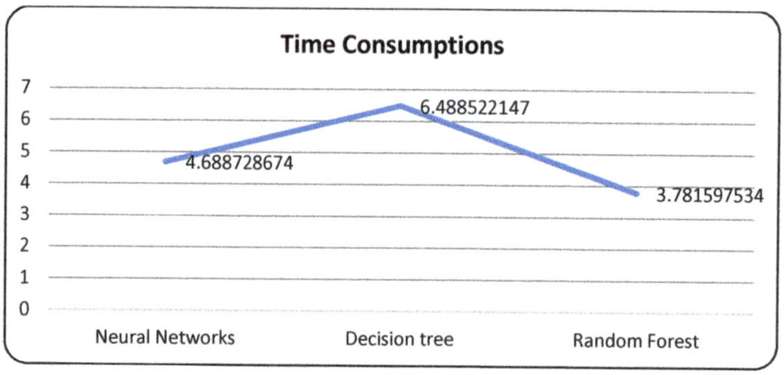

Figure 8. Models with computational time.

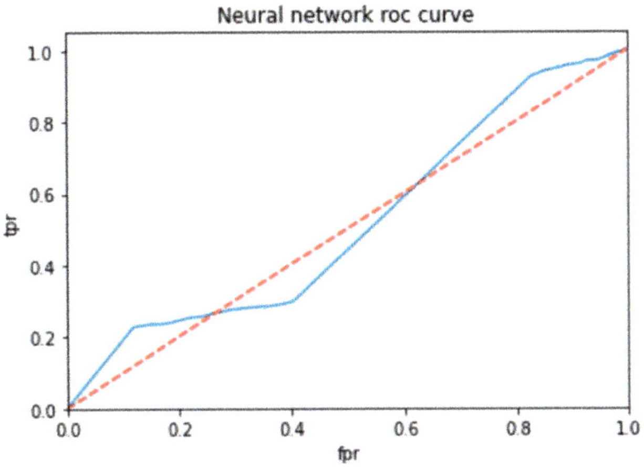

Figure 9. ROC for neural network classifier.

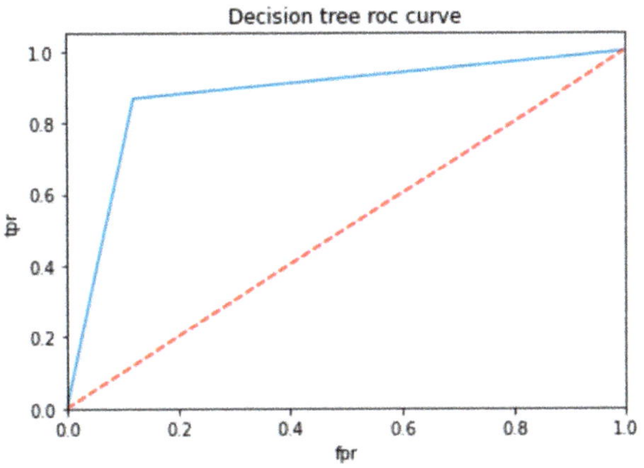

Figure 10. ROC for decision tree classifier.

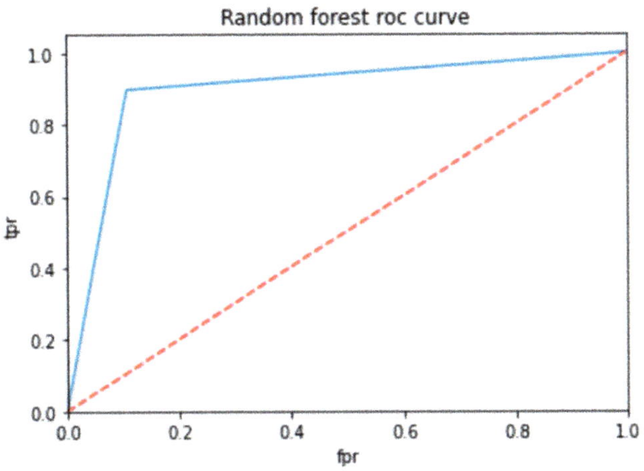

Figure 11. ROC for random forest classifier.

Figure 9 shows the optimized value as a red line and the ROC points as a blue line. To be regarded as a good model, the ROC curve must be larger than the optimized region. The model has to be improved if the region is smaller than the optimized one.

In Table 14, the AUC (area under the curve) score is a popular metric for evaluating the performance of binary classifiers. In the context of neural networks, decision trees, and random forests, the AUC score measures how well these models can distinguish between positive and negative samples. A higher AUC score indicates better classifier performance: a score of 0.5 indicates random guessing, and a score of 1.0 indicates perfect classification. Neural networks, decision trees, and random forests can all be trained to optimize the AUC score and improve their ability to classify samples correctly. As compared to other models, the proposed model is the best with a marginal difference.

Table 14. AUC scores for classifiers.

Classifiers	AUC Score
Neural Network	0.874033334
Decision Tree	0.873343619
Random Forest	0.894642082
Propose Model	0.879465732

6. Conclusions and Future Work

This study focuses on developing a predictive model for SARS-CoV-2 using machine-learning techniques, aiming to enhance prediction accuracy and provide valuable insights into the factors influencing the spread of the virus. The proposed model incorporates stochastic regression to capture the stochastic nature of virus transmission, considering uncertainties in the data. Additionally, feature selection methods are employed to identify the most relevant features contributing to prediction accuracy. The study also explores the use of neighbor embedding and Sammon mapping algorithms to visualize the high-dimensional SARS-CoV-2 data in a lower-dimensional space, facilitating better interpretation and understanding of underlying patterns. Various algorithms, including the SARS-CoV-2 dataset, combined with ADHPSRNESM, are utilized in the analysis, and a new approach called the proposed model is suggested by combining these methodologies. The research findings can contribute to public health efforts by enabling informed decision making for policymakers and healthcare professionals, ultimately aiding in effectively managing and controlling the SARS-CoV-2 pandemic. The research on enhancing accuracy through feature selection and visualization of high-dimensional data in the context of SARS-CoV-2 has substantial implications for predicting and understanding the spread of the virus. By employing various feature selection methods, the study aims to identify the most relevant and informative factors that contribute to the accuracy of SARS-CoV-2 predictions, eliminating noise and unnecessary variables for improved prediction performance. Additionally, the research incorporates neighbor embedding and Sammon mapping algorithms to visualize the high-dimensional SARS-CoV-2 data in a lower-dimensional space, enabling the identification of clusters, trends, and relationships that may not be apparent in the original data.

These findings have practical implications for public health efforts, empowering policymakers and healthcare professionals to allocate resources, implement control measures, and target interventions based on accurate predictions and a better understanding of the underlying patterns driving SARS-CoV-2 transmission. Ultimately, this research contributes to the global response in managing and controlling the SARS-CoV-2 outbreak, minimizing its impact on affected populations.

Author Contributions: S.K.R.: Original draft; S.R.: Conceptualization; M.M.E.: Methodologies; S.G.: Formal analysis.; F.K.K.: Data curation.; R.M.: resources; and D.S.K.: Investigation. All authors have read and agreed to the published version of the manuscript.

Funding: Princess Nourah bint Abdulrahman University Researchers Supporting Project number (PNURSP2023R300), Princess Nourah bint Abdulrahman University, Riyadh, Saudi Arabia.

Institutional Review Board Statement: The references in the article, as well as their correctness and appropriateness, are completely the writers' responsibility.

Informed Consent Statement: Not applicable.

Data Availability Statement: https://www.kaggle.com/datasets/plameneduardo/sarscov2-ctscan-dataset, accessed on 24 May 2023.

Acknowledgments: The authors would like to express our sincere gratitude and appreciation to the Princess Nourah bint Abdulrahman University Researchers Supporting Project (PNURSP2023R300) at Princess Nourah bint Abdulrahman University in Riyadh, Saudi Arabia.

Conflicts of Interest: The authors declare no conflict of interest.

References

1. Bhattacharya, M.; Alshammari, A.; Alharbi, M.; Dhama, K.; Lee, S.S.; Chakraborty, C. A novel mutation-proof, next-generation vaccine to fight against upcoming SARS-CoV-2 variants and subvariants, designed through AI enabled approaches and tools, along with the machine learning based immune simulation: A vaccine breakthrough. *Int. J. Biol. Macromol.* **2023**, *17*, 124893. [CrossRef]
2. Moitra, P.; Chaichi, A.; Hasan, S.M.; Dighe, K.; Alafeef, M.; Prasad, A.; Gartia, M.R.; Pan, D. Probing the mutation independent interaction of DNA probes with SARS-CoV-2 variants through a combination of surface-enhanced Raman scattering and machine learning. *Biosens. Bioelectron.* **2022**, *208*, 114200. [CrossRef]
3. Beduk, D.; de Oliveira Filho, J.I.; Beduk, T.; Harmanci, D.; Zihnioglu, F.; Cicek, C.; Sertoz, R.; Arda, B.; Goksel, T.; Turhan, K.; et al. 'All In One' SARS-CoV-2 variant recognition platform: Machine learning-enabled point of care diagnostics. *Biosens. Bioelectron. X* **2022**, *10*, 100105. [CrossRef]
4. Taft, J.M.; Weber, C.R.; Gao, B.; Ehling, R.A.; Han, J.; Frei, L.; Metcalfe, S.W.; Overath, M.D.; Yermanos, A.; Kelton, W.; et al. Deep mutational learning predicts ACE2 binding and antibody escape to combinatorial mutations in the SARS-CoV-2 receptor-binding domain. *Cell* **2022**, *185*, 4008–4022. [CrossRef]
5. Sokhansanj, B.A.; Rosen, G.L. Predicting COVID-19 disease severity from SARS-CoV-2 spike protein sequence by mixed effects machine learning. *Comput. Biol. Med.* **2022**, *149*, 105969. [CrossRef]
6. Taha, B.A.; Al Mashhadany, Y.; Al-Jubouri, Q.; Rashid, A.R.; Luo, Y.; Chen, Z.; Rustagi, S.; Chaudhary, V.; Arsad, N. Next-generation nanophotonic-enabled biosensors for intelligent diagnosis of SARS-CoV-2 variants. *Sci. Total Environ.* **2023**, *880*, 163333. [CrossRef]
7. Park, J.J.; Chen, S. Metaviromic identification of discriminative genomic features in SARS-CoV-2 using machine learning. *Patterns* **2022**, *3*, 100407. [CrossRef]
8. Budhiraja, I.; Garg, D.; Kumar, N.; Sharma, R. A comprehensive review on variants of SARS-CoVs-2: Challenges, solutions and open issues. *Comput. Commun.* **2022**, *197*, 34–51.
9. Xiao, H.; Hu, J.; Huang, C.; Feng, W.; Liu, Y.; Kumblathan, T.; Tao, J.; Xu, J.; Le, X.C.; Zhang, H. CRISPR techniques and potential for the detection and discrimination of SARS-CoV-2 variants of concern. *TrAC Trends Anal. Chem.* **2023**, *161*, 117000. [CrossRef]
10. Beguir, K.; Skwark, M.J.; Fu, Y.; Pierrot, T.; Carranza, N.L.; Laterre, A.; Kadri, I.; Korched, A.; Lowegard, A.U.; Lui, B.G.; et al. Early computational detection of potential high-risk SARS-CoV-2 variants. *Comput. Biol. Med.* **2023**, *155*, 106618. [CrossRef]
11. Philot, E.A.; Gasparini, P.; Magro, A.J.; Mattos, J.C.; Torres-Bonfim, N.E.; Kliousoff, A.; Quiroz, R.C.; Perahia, D.; Scott, A.L. Unveiling Mutation Effects on the Structural Dynamics of the Main Protease from SARS-CoV-2 with Hybrid Simulation Methods. *bioRxiv* **2021**. [CrossRef]
12. Nguyen, T.H.; Tam, N.M.; Van Tuan, M.; Zhan, P.; Vu, V.V.; Quang, D.T.; Ngo, S.T. Searching for potential inhibitors of SARS-CoV-2 main protease using supervised learning and perturbation calculations. *Chem. Phys.* **2023**, *564*, 111709. [CrossRef] [PubMed]
13. Nagpal, S.; Pinna, N.K.; Pant, N.; Singh, R.; Srivastava, D.; Mande, S.S. Can machines learn the mutation signatures of SARS-CoV-2 and enable viral-genotype guided predictive prognosis? *J. Mol. Biol.* **2022**, *434*, 167684. [CrossRef] [PubMed]
14. Fraser, D.D.; Patel, M.A.; Van Nynatten, L.R.; Martin, C.; Seney, S.L.; Miller, M.R.; Daley, M.; Slessarev, M.; Cepinskas, G.; Juneja, G.K.; et al. Cross-immunity against SARS-CoV-2 variants of concern in naturally infected critically ill COVID-19 patients. *Heliyon* **2023**, *9*, e12704. [CrossRef]
15. Singh, P.; Ujjainiya, R.; Prakash, S.; Naushin, S.; Sardana, V.; Bhatheja, N.; Singh, A.P.; Barman, J.; Kumar, K.; Gayali, S.; et al. A machine learning-based approach to determine infection status in recipients of BBV152 (Covaxin) whole-virion inactivated SARS-CoV-2 vaccine for serological surveys. *Comput. Biol. Med.* **2022**, *146*, 105419. [CrossRef] [PubMed]

16. Bucinsky, L.; Bortňák, D.; Gall, M.; Matúška, J.; Milata, V.; Pitoňák, M.; Štekláč, M.; Végh, D.; Zajaček, D. Machine learning prediction of 3CLpro SARS-CoV-2 docking scores. *Comput. Biol. Chem.* **2022**, *98*, 107656. [CrossRef]
17. Kc, G.B.; Bocci, G.; Verma, S.; Hassan, M.M.; Holmes, J.; Yang, J.J.; Sirimulla, S.; Oprea, T.I. A machine learning platform to estimate anti-SARS-CoV-2 activities. *Nat. Mach. Intell.* **2021**, *3*, 527–535. [CrossRef]
18. Du, R.; Tsougenis, E.D.; Ho, J.W.; Chan, J.K.; Chiu, K.W.; Fang, B.X.; Ng, M.Y.; Leung, S.T.; Lo, C.S.; Wong, H.Y.; et al. Machine learning application for the prediction of SARS-CoV-2 infection using blood tests and chest radiograph. *Sci. Rep.* **2021**, *11*, 14250. [CrossRef]
19. Rauseo, M.; Perrini, M.; Gallo, C.; Mirabella, L.; Mariano, K.; Ferrara, G.; Santoro, F.; Tullo, L.; La Bella, D.; Vetuschi, P.; et al. Machine learning and predictive models: 2 years of SARS-CoV-2 pandemic in a single-center retrospective analysis. *J. Anesth. Analg. Crit. Care* **2022**, *2*, 42. [CrossRef]
20. Cihan, P.; Ozger, Z.B. A new approach for determining SARS-CoV-2 epitopes using machine learning-based in silico methods. *Comput. Biol. Chem.* **2022**, *98*, 107688. [CrossRef]
21. De Oliveira, R.F.; Bastos Filho, C.J.; de Medeiros, A.C.; dos Santos, P.J.; Freire, D.L. Machine learning applied in SARS-CoV-2 COVID-19 screening using clinical analysis parameters. *IEEE Lat. Am. Trans.* **2021**, *19*, 978–985. [CrossRef]
22. Bishnu, S.K.; Das, R.; Ganguly, R.; Biswas, R.; Bose, R.; Dhar, S.; Ray, S. Diagnostic Mask for SARS-CoV-2 Detection Using Immunochromatography and Machine Learning. In Proceedings of the 2021 5th International Conference on Electronics, Materials Engineering & Nano-Technology (IEMENTech), Kolkata, India, 24–26 September 2021; IEEE: Piscataway, NJ, USA, 2021; pp. 1–6.
23. Begum, F.; Mridha, K.; Rabbani, M.G.; Ashfaq, S.; Islam, S.M.; Sinha, S. Bioactivity Classification of SARS-CoV-2 Proteinase Using Machine Learning Approaches. In Proceedings of the 2022 10th International Conference on Reliability, Infocom Technologies and Optimization (Trends and Future Directions) (ICRITO), Noida, India, 13–14 October 2022; IEEE: Piscataway, NJ, USA, 2022; pp. 1–6.
24. Togrul, M.; Arslan, H. Detection of SARS-CoV-2 Main Variants of Concerns using Deep Learning. In Proceedings of the 2022 Innovations in Intelligent Systems and Applications Conference (ASYU), Antalya, Turkey, 7–9 September 2022; IEEE: Piscataway, NJ, USA, 2022; pp. 1–5.
25. Hu, F.; Wang, D.; Hu, Y.; Jiang, J.; Yin, P. Generating novel compounds targeting SARS-CoV-2 main protease based on imbalanced dataset. In Proceedings of the 2020 IEEE International Conference on Bioinformatics and Biomedicine (BIBM), Seoul, Republic of Korea, 16–19 December 2020; IEEE: Piscataway, NJ, USA, 2020; pp. 432–436.
26. Jain, S.; Kumar, M.; Sindhwani, N.; Singh, P. SARS-CoV-2 detection using Deep Learning Techniques on the basis of Clinical Reports. In Proceedings of the 2021 9th International Conference on Reliability, Infocom Technologies and Optimization (Trends and Future Directions) (ICRITO), Noida, India, 3–4 September 2021; IEEE: Piscataway, NJ, USA, 2021; pp. 1–5.
27. Mohamed, T.; Sayed, S.; Salah, A.; Houssein, E.H. Next generation sequence prediction intelligent system for SARS-CoV-2 using deep learning neural network. In Proceedings of the 2021 17th International Computer Engineering Conference (ICENCO), Cairo, Egypt, 29–30 December 2021; IEEE: Piscataway, NJ, USA, 2021; pp. 88–93.
28. Moosavi, J.; Fathollahi-Fard, A.M.; Dulebenets, M.A. Supply chain disruption during the COVID-19 pandemic: Recognizing potential disruption management strategies. *Int. J. Disaster Risk Reduct.* **2022**, *75*, 102983. [CrossRef] [PubMed]
29. Fathollahi-Fard, A.M.; Ahmadi, A.; Karimi, B. Sustainable and robust home healthcare logistics: A response to the COVID-19 pandemic. *Symmetry* **2022**, *14*, 193. [CrossRef]

bioengineering

MDPI

Article

Numerical Modeling and Simulation of Non-Invasive Acupuncture Therapy Utilizing Near-Infrared Light-Emitting Diode

Sundeep Singh, Andres Escobar, Zexi Wang, Zhiyi Zhang, Chundra Ramful and Chang-Qing Xu

[1] Faculty of Sustainable Design Engineering, University of Prince Edward Island, Charlottetown, PE C1A 4P3, Canada
[2] Department of Biomedical Engineering, McMaster University, Hamilton, ON L8S 4L8, Canada; escoba3@mcmaster.ca
[3] Department of Engineering Physics, McMaster University, Hamilton, ON L8S 4L8, Canada; wangz418@mcmaster.ca
[4] Advanced Electronics and Photonics Research Center, National Research Council Canada, Ottawa, ON K1A 0R6, Canada; zhiyi.zhang@nrc-cnrc.gc.ca (Z.Z.); chundra.ramful@nrc-cnrc.gc.ca (C.R.)
* Correspondence: sunsingh@upei.ca (S.S.); cqxu@mcmaster.ca (C.-Q.X.)

Abstract: Acupuncture is one of the most extensively used complementary and alternative medicine therapies worldwide. In this study, we explore the use of near-infrared light-emitting diodes (LEDs) to provide acupuncture-like physical stimulus to the skin tissue, but in a completely non-invasive way. A computational modeling framework has been developed to investigate the light-tissue interaction within a three-dimensional multi-layer model of skin tissue. Finite element-based analysis has been conducted, to obtain the spatiotemporal temperature distribution within the skin tissue, by solving Pennes' bioheat transfer equation, coupled with the Beer-Lambert law. The irradiation profile of the LED has been experimentally characterized and imposed in the numerical model. The experimental validation of the developed model has been conducted through comparing the numerical model predictions with those obtained experimentally on the agar phantom. The effects of the LED power, treatment duration, LED distance from the skin surface, and usage of multiple LEDs on the temperature distribution attained within the skin tissue have been systematically investigated, highlighting the safe operating power of the selected LEDs. The presented information about the spatiotemporal temperature distribution, and critical factors affecting it, would assist in better optimizing the desired thermal dosage, thereby enabling a safe and effective LED-based photothermal therapy.

Keywords: bio-heat transfer; light-tissue interaction; photothermal effect; acupuncture; acupoints; thermal therapy; computational modeling

Citation: Sundeep Singh, Andres Escobar, Zexi Wang, Zhiyi Zhang, Chundra Ramful and Chang-Qing Xu Numerical Modeling and Simulation of Non-Invasive Acupuncture Therapy Utilizing Near-Infrared Light-Emitting Diode. *Bioengineering* **2023**, *10*, 837. https://doi.org/10.3390/bioengineering10070837

Academic Editors: Enrique Berjano and Luca Mesin

Received: 26 June 2023
Revised: 10 July 2023
Accepted: 13 July 2023
Published: 15 July 2023

1. Introduction

Over the past two decades, there has been a significant increase in the use of complementary and alternative medicine worldwide, across all age groups [1–7]. In particular, complementary and alternative medicine includes diverse and abundant traditional practices, such as naturopathy, chiropractic care, acupuncture, Ayurveda, and homeopathy. Among these therapies, acupuncture has received significant attention, and is one of North America's fastest-growing complementary and alternative medicine therapies [8]. Acupuncture is a nonpharmacological and minimally invasive traditional Chinese medicine therapy that originated more than 2000 years ago, and has gradually spread and become quite popular in the West [9,10]. Significant research efforts have been devoted in the past four decades to investigating acupuncture's clinical effectiveness and mechanisms of action, with more than 10,000 randomized controlled trials published since 1975 [10–13]. However, the exact explanation for the biological mechanism of action involved in acupuncture still remains elusive, which has hindered its modern development and integration into conventional medicine [14].

Today, acupuncture is widely used to treat a plethora of diseases, such as musculoskeletal pain, gastrointestinal disorders, stroke, gynecological diseases, neurological disorders, and various types of incurable chronic disease [1,15–18]. During modern acupuncture therapy, very thin needles are inserted into the skin at certain specific acupoints in the human body. The acupuncture needles are then stimulated by applying mechanical, electrical, or other physical forms of energy, to activate various nerve receptors directly or indirectly [17,19]. Several hypotheses have been proposed to describe the molecular pathways associated with the therapeutic effects of acupoint stimulations. One such recent hypothesis is related to the activation of mast cells during acupuncture, which leads to analgesic effects [14]. Moreover, the distinction between acupoints and non-acupoints is made based on the concentration of mast cells. As such, acupoints have a higher concentration of mast cells than any other location. The stimulation of the acupoints during acupuncture therapy results in the degranulation of mast cells, which further modulates neighboring cell behaviors by releasing biological substances, thereby enhancing vascular permeability and nerve activation [14].

In the broader context of acupuncture therapy, the stimulation of the acupoints can be performed through mechanical (acupuncture), heat (moxibustion), and electrical (electroacupuncture) treatments. The present study focuses on the stimulation of acupoints with heat, also called moxibustion therapy. As such, moxibustion is a non-invasive thermal therapy that completely omits the need for needle insertion, thereby mitigating any chance of pain, infection, or damage to the skin layer. Moxibustion is a relatively simple and tolerable treatment, whereby dried mugwort (moxa) is burned over the acupoint, either directly touching the skin or at a certain distance (or gap) from the skin surface [20,21]. Existing research suggests that the high temperature (approximately 500 °C) of burning moxa sticks could result in a temperature rise within the skin tissue ranging between 38 °C and 50 °C [22–24]. However, the safe and effective temperature to avoid the patient's discomfort and irreversible damage has been reported to be above 42 °C and below 50 °C [24,25]. Furthermore, moxibustion is typically performed for 10–30 min, repeated several times in a week for a few weeks, and is able to provide a thermal stimulus to the depth of 2–4 mm, affecting both the shallow and deep skin tissue [22,23]. The therapeutic effects induced during moxibustion are related to a combination of heat effects, nonthermal radiation effects, and pharmacological effects [22,24,26,27]. Among these, heat stimulation is believed to be one of the critical factors directly related to the efficacy of moxibustion therapy.

Furthermore, the major disadvantage of moxibustion is related to producing harmful gases and odors by burning moxa sticks, similar to tobacco smoke and air pollutants [23,25]. Several alternatives have been explored to address this disadvantage, and to attain similar therapeutic effects related to thermal stimulation during moxibustion therapy but without the actual production of smoke or odor. More recently, the applications of lasers and ultrasonic devices have also been explored to replace moxa sticks and attain a similar rise in temperature within the deep tissues to that achieved in moxibustion therapy [23,25,28,29]. Moving in this direction, in the present study, we aim to propose and evaluate the feasibility of moxibustion-like treatment utilizing near-infrared LED to provide thermal stimulus to the tissue in a completely non-invasive fashion.

Compared to the lasers typically used in complicated clinical settings, LEDs are safer and more cost-effective, and are well-tolerated by patients of all age groups. Furthermore, there are a plethora of visible blue and red light wavelength LED-based devices already available on the market, used by either patients or healthy individuals, independently in their homes, for some generic dermatological applications [30–32]. However, to the best of the authors' knowledge, no studies (or devices) are available in the literature that explore the feasibility of near-infrared LEDs as a reasonable alternative to moxa sticks during moxibustion therapy, and that quantify the relationship between the light-tissue interactions during such a treatment. The selection of near-infrared LEDs is also based on the fact that it reaches the maximum depth in the skin. In the following sections, we first characterize the light irradiation of selected LEDs on the surface of the tissue. Then, we utilize a numerical modeling framework to quantify the temperature rise within the skin tissue subjected to the Gaussian irradiation profile of the LED. Furthermore, parametric studies have been conducted, to investigate the effect of the applied power, irradiation duration, and LED distance from the skin surface on the temperature rise attained at different depths in the skin tissue. The results of the parametric analysis reported in the present study will be helpful in better understanding the effects of various extrinsic factors on the LED-based heating of skin tissue. They would pave the way to developing cost-effective and easy-to-use LED-based wearable devices, to provide on-demand physical stimulus at home, without needing to go to a clinic.

2. Materials and Methods

This section provides details of the computational domain and numerical framework adopted to study the effect of LED irradiation on heat transfer in skin tissue. The optical and thermal properties of the tissue, governing equations, and boundary conditions are also presented. Finally, the details of the characterization of the irradiance profile for the selected LEDs, and the experimental setup utilized to test and validate the developed numerical model, have been provided.

2.1. Finite-Element-Method (FEM)-Based Model

Figure 1 presents the computational domain of skin tissue considered in the present study. The skin model comprises three layers: the epidermis, the dermis, and subcutaneous tissue [33–35]. Each layer has distinct optical and thermal properties. LEDs were placed on the top surface of the skin tissue, with variable distances between the top of the LED surface and the epidermis skin layer. The main motive of the present study was to quantify the interaction between the LED irradiance and the temperature rise within the skin tissue. In addition, heat transfer analysis within biological tissue is highly complex, due to the heterogeneous vascular structure, blood flow in the complex network of arteries and veins, metabolic heat generation, and dependence of the tissue properties on the physiological conditions [36–38]. Pennes' Fourier-law-based bioheat transfer model is widely used for predicting the temperature distribution within the human body for numerous biological and medical applications, due to its simplicity, computational efficiency, and cost-effectiveness. Notably, Pennes' model is a continuum model, whereby the blood vessels are not individually incorporated into the computational domain; instead, their effect is lumped into a single factor known as the "blood perfusion rate".

The generalized version of Pennes' bioheat transfer equation utilized in the present study for evaluating the LED-induced heating is represented as [38]:

$$\rho c_p \frac{\partial T}{\partial t} = \nabla \cdot (k \nabla T) - \rho_b c_b \omega_b (T - T_b) + Q_{met} + Q_{source} \tag{1}$$

where ρ is the density of the tissue (kg/m^3); c_p is the specific heat capacity of the tissue (J/kg/K); k is the thermal conductivity of the tissue (W/m/K); T is the temperature of the tissue that will be quantified from Equation (1) (K); ρ_b is the density of the blood (=1050 kg/m^3) [39]; c_b is the specific heat of the blood (=3617 J/kg/K) [39]; ω_b is the blood perfusion rate (s^{-1}); T_b is the arterial blood temperature (=310 K); Q_{met} is the metabolic heat generation (=368 W/m^3) [34,35]; and Q_{source} represents the energy absorption of the LED irradiation (W/m^3).

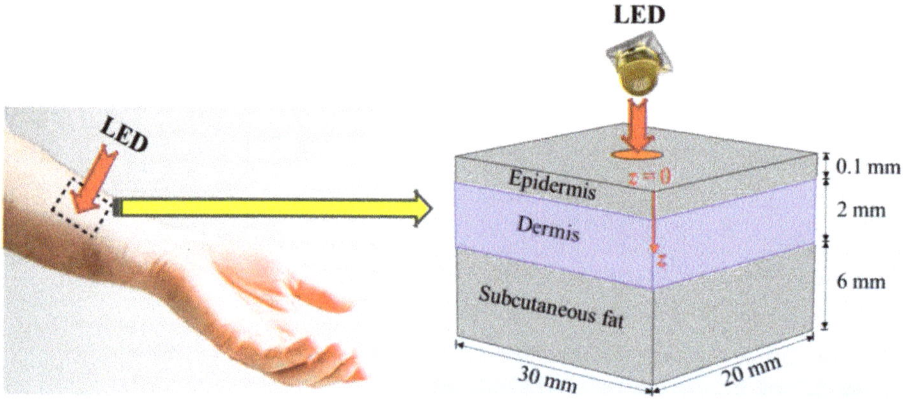

Figure 1. Schematic of the 3D model of the three-layered skin considered in the present study.

The LED heat source has been modeled using the Beer-Lambert law in the numerical analysis, which is represented as [34]:

$$Q_{source} = \mu_{eff} \cdot I \cdot e^{-\mu_{eff} \cdot z} \quad (2)$$

where μ_{eff} is the effective attenuation coefficient (m^{-1}); z is the depth of the tissue (m); and I is the LED irradiation intensity (W/m^2) given by [40]:

$$I(x,y) = I_o \cdot e^{-\frac{(x^2+y^2)}{2\sigma^2}} = \frac{P}{2\pi\sigma^2} \cdot e^{-\frac{(x^2+y^2)}{2\sigma^2}} \quad (3)$$

where I_0 is the maximum irradiation intensity at the center of the 2D Gaussian profile of the LED beam (W/m^2); σ is the standard deviation of the LED beam Gaussian profile; and P is the LED radiometric power (W). As mentioned, the LED operates at a near-infrared wavelength of 850 nm. Furthermore, the effective attenuation coefficient that accounts for the absorption and scattering phenomena of LED light in skin tissue, based on diffusion approximation, is given as [34]:

$$\mu_{eff} = \sqrt{3\mu_a(\mu_a + \mu'_s)} \quad (4)$$

where μ_a is the absorption coefficient of tissue (m^{-1}), and μ'_s is the reduced scattering coefficient of the tissue (m^{-1}), both approximated at a wavelength of 850 nm for different skin layers, from [41]. The thermal and optical properties of the different skin layers (the epidermis, dermis, and subcutaneous fat) considered in the numerical simulations are summarized in Table 1 [34,35]. More details of the characterization of material properties utilizing photothermoacoustic spectroscopy can be found in [42].

Table 1. Thermal and optical properties of the various skin layers considered in the present study.

Parameters	Epidermis	Dermis	Subcutaneous Fat
Density, ρ (kg/m^3)	1200	1090	1210
Specific heat capacity, c_p (J/kg/K)	3950	3350	2240
Thermal conductivity, k (W/m·K)	0.24	0.42	0.194
Blood perfusion rate, ω_b (1/s)	-	0.002	0.002
Absorption coefficient @ 850 nm, μ_a (1/cm)	0.9	0.95	1
Reduced scattering coefficient @ 850 nm, μ'_s (1/cm)	30	20	18

FEM-based COMSOL Multiphysics 6 (COMSOL, Inc., Burlington, MA, USA) software was adopted to solve the spatiotemporal temperature distribution obtained within the skin tissue subjected to LED irradiation. The computational domain was discretized using unstructured tetrahedral mesh elements in COMSOL. A grid independence solution was obtained by refining the mesh until the error associated with the maximum temperature prediction was less than 0.5% between two contiguous meshes. The optimal number of total mesh elements considered in the present study was 162,118. The initial temperature of the skin tissue was assumed to be the same as the core body temperature of the human body, 37 °C. The Gaussian profile obtained from experimental characterization was applied at the top surface of the computational domain of the skin, at the center, for 10 min. All the computational domain boundaries, except for the top, were applied with thermal insulated boundary conditions. The top surface of the computational domain was subjected to the natural convection and radiation boundary conditions given by $(-k\nabla T = h(T - T_{\infty}))$, and $(-k\nabla T = \varepsilon\sigma(T^4 - T_{\infty}^4))$, respectively, where h is the convective heat transfer coefficient assumed to be 10 W/m^2 K [23,35]; T_{∞} is the surrounding ambient temperature, considered to be 25 °C [23,35]; ε is the emissivity of the skin surface, assumed to be 0.95 [23,34,35]; and σ is the Stefan-Boltzmann constant.

2.2. LED Irradiance Profile

The characterization of the optical properties of the selected near-infrared LEDs (LUXEON IR Domed Line, Lumileds Holding B.V.) operating at a wavelength of 850 nm, with 90° and 150° emission patterns representing full-width half maximum (FWHM) angles, was performed utilizing a Thorlabs BP209IR1 beam profiler, as shown in Figure 2a. The XYZ stage was utilized to make micron-level adjustments that would facilitate the alignment of the LED with the center of the beam profiler, to ensure a properly characterized beam spread. This XYZ stage was fitted with a rod that allowed a clamp holding the LED-heat-sink combination to move freely. Once aligned, the LED was moved as close to the aperture as possible, and its position was recorded on the XYZ stage as a relative position zero. Having the LED measured physically close to the aperture would help increase the accuracy of the beam profile shape, its peak intensity, and other optical features, as the entirety of the light produced would be captured within the aperture. With this setup, we could then take beam profile measurements at different relative distances, to better characterize the beam shape and how it would change at increasing distances, to be used as the input in our developed numerical models for predicting the spatiotemporal temperature distribution within the biological tissue. Before any measurements were taken, the beam profile was set to accept the LED operating wavelength of 850 nm, as well as being set to operate at the maximum aperture width of 9.0 mm, and at a scanning rate of 20 Hz. With the increased scanning rate, the resolution of our profiles was increased, and helped to reduce the warping effect of the potential saturation incurred by the LED on the beam profiler. Customized software and a customized circuit board were used in the experiments. The LED profile for both LEDs was represented by a Gaussian-fit curve, as represented in Figure 2b. This Gaussian irradiation profile for both LEDs was obtained for different operating currents and voltages, and thus different operating power. The obtained Gaussian irradiance profile was provided as an input at the top of the skin surface in the numerical model for quantifying the light-tissue interaction. The obtained data were in agreement with the manufacturer's specifications and reported values.

2.3. Experimental Setup for Testing the LEDs

Agar gel was prepared with a 2.6% (w/v) concentration of agar powder mixed in DI water. The recipe for the agar gel was adopted from [43], owing to the availability of optical and thermal properties at that concentration. The fabricated agar gel was cylindrical, with a height of 0.9 cm and a diameter of 7.5 cm. Figure 3 presents the experimental setup utilized for testing the LEDs on the agar gel. The LED was applied at the top surface of the agar gel, and a type-K thermocouple was gently inserted in the agar gel, to keep the monitoring tip at a distance of 2 mm just below the center of the LED from the top surface. For experimental validation, three sets of trials (for each LED) were conducted, to quantify the temperature profile at a depth of 2 mm within the agar tissue exposed to the 90° and 150° LEDs operated at a current of 0.802 A and a voltage of 3.6 V, for 5 min.

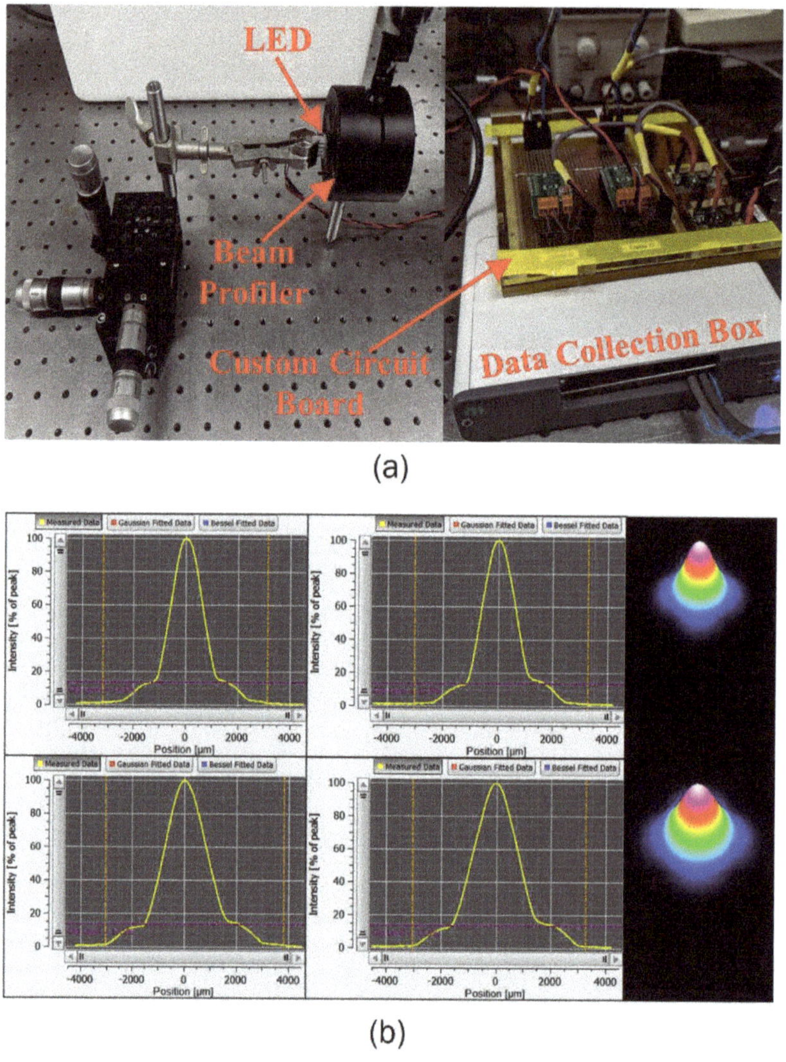

Figure 2. (**a**) The experimental setup used to determine the LED beam profiles. The different components of the experimental setup are the XYZ stage, LED heatsink, beam profiler, and a custom-made circuit board for operating the LEDs. (**b**) A snapshot of a typical beam-profiling measurement. From left to right: the 2D *x*-axis, 2D *y*-axis, and 3D beam profiles of the 90° (top row) and 150° LEDs (bottom row).

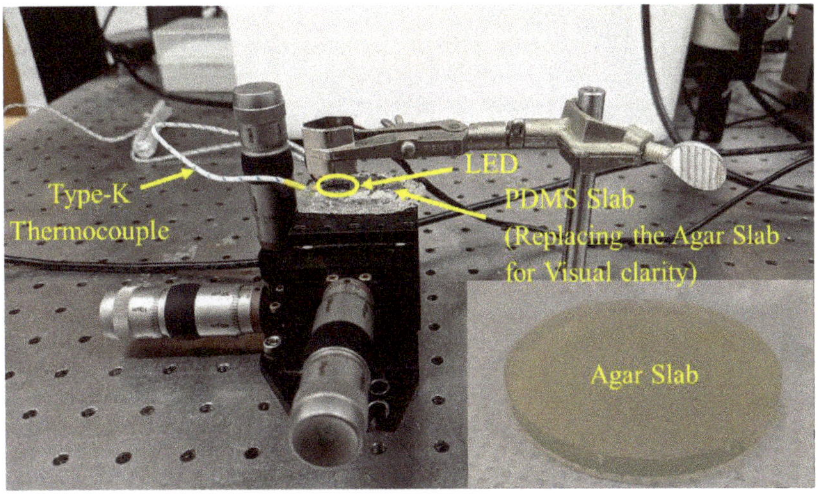

Figure 3. Photographic view of the experimental setup for measuring the temperature profile. The snapshot in the bottom right-hand corner is a sample of the actual agar gel used in the temperature profile experiments.

3. Results

3.1. Experimental Validation

The developed model's fidelity and integrity have been evaluated by comparing the results predicted from the numerical simulation with those obtained numerically from experimental studies on agar phantom under the same environmental conditions. Notably, the temperature profile obtained at the location of 2 mm away from the center of the LED along the tissue depth was used for validation purposes. Motivated by [43], the thermal and optical properties of the agar phantom were considered to be ρ = 1050 kg/m^3, c_p = 4219 J/kg/K, k = 0.66 W/m/K, μ_a = 40 m^{-1}, and μ'_s = 530 m^{-1}. The effects of blood perfusion and metabolic heat generation were neglected in the numerical model, to have consistency with the agar phantom. The comparison of the temperature profile attained with 5 min of LED irradiation has been presented in Figure 4. The mean standard deviation along the whole trend of the experimental tests is 3.19% and 2.1%, for the 90° and 150° emission pattern LEDs, respectively, thus proving a limited variability across the experiments. The correlation factor, R, between the numerically predicted and experimental data after 300 s of LED irradiation is found to be 0.98 and 0.97, for the 90° and 150° emission pattern LEDs, respectively. The mean absolute error is found to be 0.95 °C and 1 °C, for the 90° and 150° emission pattern LEDs, respectively, which shows a good performance. As shown in Figure 4, there is a good agreement between the temperature profile obtained with the numerical simulation and with experiments, for both the 90° and 150° emission pattern LEDs, thereby lending great confidence to the results reported and discussed in the following sections. Next, we will present the results of the numerical model, extended to incorporate the optical and thermal properties of skin tissue, including microvascular blood perfusion.

3.2. Effect of Input LED Power on Temperature Rise within the Skin

In this section, we will report the results of the parametric analysis conducted to quantify the effect of the input LED power and the duration on the temperature rise attained within the skin tissue. The treatment time for the therapy was selected to be 10 min to attain thermal stimulus (>43 °C) at the depth $\geq$ 2 mm, consistent with typical moxibustion therapy. The parametric analysis was conducted on the two LEDs operating at a near-infrared wavelength of 850 nm, with 90° and 150° emission patterns representing full-width half maximum (FWHM) angles. Although these LEDs can be operated at the maximum power of 1000 mW, to avoid any heating and self-damaging issues, the maximum operating power of the LED was kept as 700 mW. Figure 5a presents the transient temperature increase obtained within the skin tissue when the 90° LED was placed at the top of the three-layer skin model, with direct contact. As evident from Figure 5a, the increase in the operating power of the LED resulted in a corresponding increase in the maximum temperature rise within the skin tissue. The relative jump in maximum temperature attained by increasing the LED power by a factor of 100 mW was approximately the same for all the cases. Further, the maximum temperature reached within the skin tissue was less than 50 °C for the LED operating at 300 mW, beyond which the maximum temperature was greater than 50 °C. For all the LED operating powers, initially, the temperature rose steeply and later stabilized and saturated with time. The higher the LED's operating power, the longer the time required to attain this saturation. For all considered cases, the saturation was attained within 5–6 min of LED therapy. This highlights that the maximum temperature will be constant and will not increase with an increase in treatment time, i.e., the maximum temperature attained after 10 min will be the same after 15 min or 30 min, or even after one hour of LED therapy. However, the zone of the skin tissue experiencing thermal effects may still enhance with time, due to the conduction of heat deposited within the tissue.

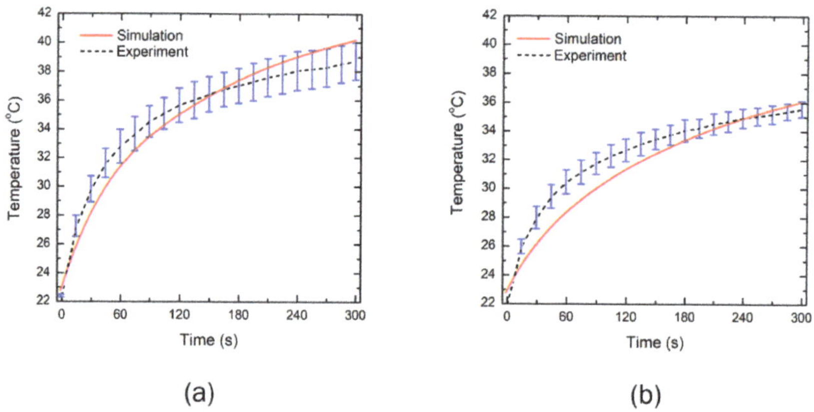

Figure 4. Comparison of the experimentally measured and numerically predicted temperature profiles at a depth of 2 mm for LEDs with emission patterns of: (**a**) 90°, and (**b**) 150°.

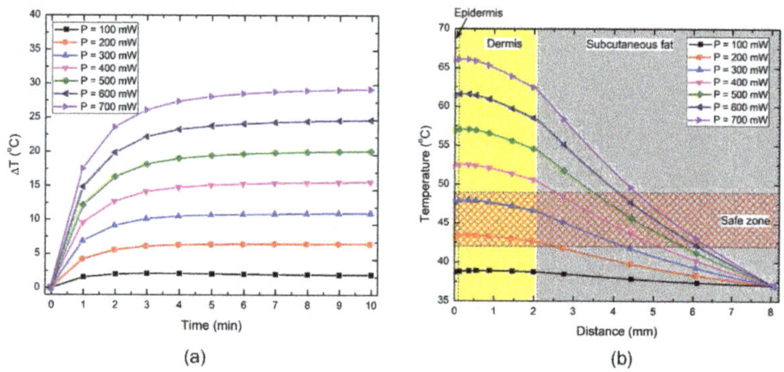

Figure 5. The effect of an input power of 90° emission pattern LED on (**a**) the maximum temperature rise with time within the skin tissue, and (**b**) the temperature variation along the depth of the skin after 10 min of LED irradiation.

Figure 5b presents the temperature distribution attained after 10 min of LED therapy within the skin tissue, as a function of the depth from the skin's outermost layer and from the LED's center. As evident from Figure 5b, the maximum temperature was not attained at the outer epidermis layer of skin; indeed, it occurred at the dermis layer. This can be attributed to the natural convection cooling effects prevalent at the outer epidermis layer, lowering the temperature slightly. After attaining the peak, the temperature continuously declined with the depth of skin tissue. Furthermore, the increased operating power of the LED increased both the temperature and the depth to which the higher-temperature effects were felt within the skin tissue. Notably, previous physiological studies have highlighted the activation of cutaneous thermosensitive receptors (TRPV channels) with increased tissue temperature, inducing different biological and chemical reactions in the human body [22,23,26,27,44]. For example, TRPV1 and TRPV2 can be activated when the heated tissue temperature exceeds 43 °C and 53 °C, respectively. Moreover, the rise in temperature at 44.5–46.5 °C can activate A-fibre mechano-heat-sensitive nociceptors in the human skin [45]. The cholesterol-lowering effect is far more pronounced at 46 °C than at 38 °C [46]. Moreover, attaining a temperature above 50 °C will result in thermal ablation, i.e., inducing irreversible damage within the skin tissue. Thus, the safe working limit of the proposed LED-based photothermal therapy would be in the temperature range of 43–48 °C, similar to the safe and effective range of moxibustion therapy. Based on the parametric studies conducted on the 90° LED, the safe operating power resulting in therapeutic effects has been found to be in the range of 200–300 mW, as highlighted in Figure 5b.

The effect of operating power of 150° LED on the maximum temperature rise has been presented in Figure 6a. Again, as is evident, the increase in the operating power of the LED resulted in an increase in the maximum temperature rise within the skin tissue, which was somewhat proportionate to the increase in power. The temperature distribution obtained within the skin tissue after 10 min of LED therapy is presented in Figure 6b. The safe operating range of power to obtain therapeutic effects with 150° LED-based photothermal therapy has been found to be 300–500 mW, as shown in Figure 6b.

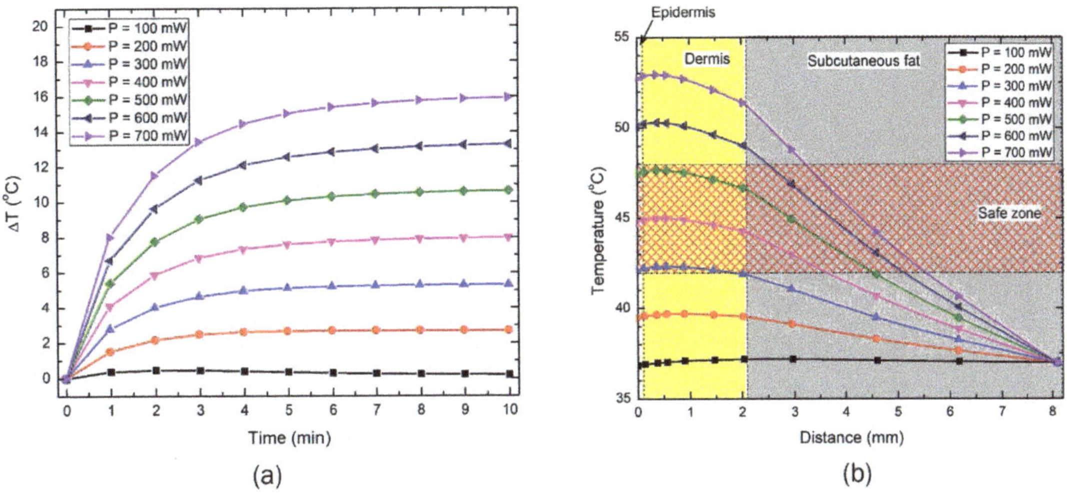

Figure 6. The effect of an input power of 150° emission pattern LED on (**a**) the maximum temperature rise with time within the skin tissue, and (**b**) the temperature variation along the depth of the skin after 10 min of LED irradiation.

Furthermore, Figure 7 presents the heated volume of the skin tissue after 10 min of LED irradiation, quantified with the isotherms of 43 °C and 50 °C. The goal of the proposed LED therapy is to maximize tissue volume attaining a temperature greater than 43 °C, and mitigate any chances of reaching 50 °C anywhere within the skin tissue. The heated volume obtained from the parametric analysis, as presented for the 90° (in Figure 7a) and 150° (in Figure 7b) emission patterns, will assist in selecting the safe operating power range, to avoid the temperature that initiates the ablation of the tissue. As depicted in Figure 7, the heated volume abruptly increased with the increased operating power of the LED. The initiation of the ablation zone (T > 50 °C) occurred at the power of 400 mW and 600 mW, for the LEDs with emission patterns of 90° and 150°, respectively. This highlights that the safe operating power for 90° and 150° should be less than 400 mW and 600 mW, respectively, to avoid the occurrence of thermal ablation leading to irreversible damage to the skin tissue.

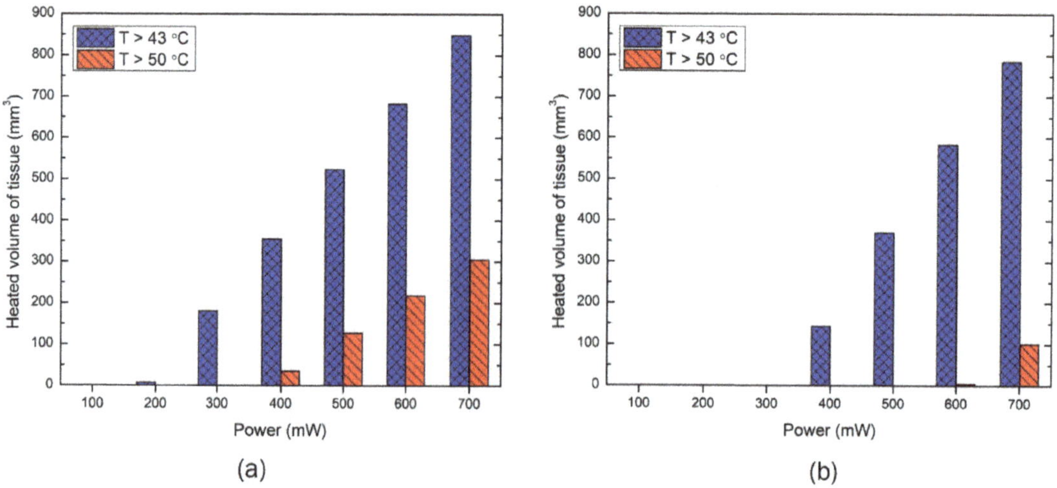

Figure 7. The variation in the heated volume obtained using temperatures greater than the 43 °C and 50 °C protocol after 10 min of LED irradiation with emission patterns of (**a**) 90°, and (**b**) 150°.

Figure 8 directly compares the two selected LEDs, with emission patterns of 90° and 150°, operating at a power of 500 mW. As evident from Figure 8a, the maximum temperature rise attained within the skin tissue is significantly higher for the 90° LED than for the 150° LED. The temperature distribution obtained within the skin tissue after 10 min of LED therapy is presented in Figure 8b, again highlighting the performance of the 90° LED being far superior to that of the 150° LED. This can be attributed to the fact that the emission angle of the 90° LED was more focused and thus could deposit a higher amount of energy in a focused area, compared to the 150° LED that had a higher divergence angle and, thus the light intensity could not penetrate to greater depths. The standard deviation of the Gaussian profile (σ) for the 90° and 150° LEDs was found to be 4.6 mm and 7 mm, respectively, highlighting the irradiation area of the 150° LED being higher, compared to the 90° LED. Therefore, if the goal is to deliver the heat to a greater depth, one would use a 90° LED, and if the goal is to heat up a large area at a lower penetration depth, one would use a 150° LED. The isotherms of 43 °C obtained after 10 min of LED-based photothermal therapy are presented in Figure 8c. As depicted in this figure, the depth to which the effect of the 43 °C temperature is felt is 5.4 mm for the 90° LED, compared to 4 mm for the 150° LED. Thus, the selection of the LED based on both the operating power and emission angle basis would significantly affect the spatiotemporal temperature distribution, and hence the efficacy of LED-based therapy.

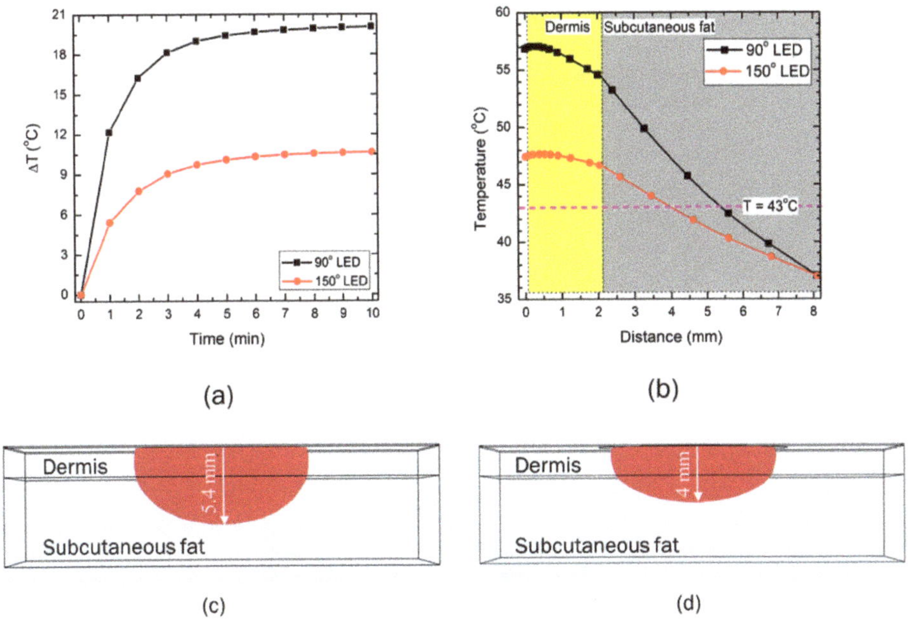

Figure 8. The effect of the LED emission patterns of 90° and 150° operating at an input power of 500 mW on (**a**) the maximum temperature rise within the skin tissue, and (**b**) the temperature distribution along the depth of the skin tissue. Heating zone obtained with the isotherm of 43 °C after 10 min of LED irradiation with the emission pattern of (**c**) 90°, and (**d**) 150°.

3.3. Effect of Distance between the LED and Skin Surface on Heat Transport

The effect of different distances between the LED and the skin's topmost layer is presented in Figures 9 and 10 for the 90° and 150° emission patterns, respectively. The LEDs were operated at the safe operating power of 300 mW and 500 mW for the 90° and 150° emissions, respectively. Two distances were selected for this analysis: (i) no gap between the LED and the skin (0 μm distance), and (ii) a gap of 2000 μm between the LED and the skin surface. Figure 9a presents the temporal variation in the maximum temperature for the two selected distances. As evident, a significant drop in the maximum temperature attained within the skin tissue prevailed, as the distance between the LED and the skin surface increased. After 10 min of LED therapy, the maximum temperature dropped by 37.6%, as the LED initially placed in direct contact with the skin surface was moved at a distance of 2000 μm from the skin surface. The temperature distribution attained along the depth of skin tissue after 10 min of therapy with the two selected locations for the 90° LED is presented in Figure 9b. The increase in the distance between the LED and the skin surface resulted in a drop in temperature at (a) the epidermis-dermis interface by 8.7%, and (b) the dermis-subcutaneous fat interface by 7.3%. This can be attributed to the fact that as the distance between the LED and the skin surface was increased, the LED irradiation intensity was no longer focused, and the losses were significant, thus resulting in an abrupt decline in the temperature attained within the skin tissue. For example, the standard deviation of the 90° LED beam Gaussian profile (σ) increased from 4.6 mm to 6.1 mm as the LED was moved at the distance of 2000 μm from the skin surface, thus leading to a decline in the irradiation intensity (as σ is inversely proportional to the irradiation intensity). The temporal variation in the volume obtained using the isotherm of 43 °C (i.e., the volume of skin tissue within the computational domain having temperature $\geq$ 43 °C) is shown in Figure 9c. As depicted in this figure, the temperature of 43 °C was attained almost instantaneously when the LED was placed in direct contact with the skin surface. In comparison, it took almost 3 min before the temperature of 43 °C was attained for the LED placed at a distance of 2000 µm from the skin surface. After 10 min of LED-based heating of the skin tissue, the desired heated volume of the tissue dropped abruptly by 372%, as the distance between the LED and the topmost layer of skin was increased to 2000 µm. Furthermore, the temperature distribution attained after 10 min of LED heating in the three-layered skin tissue is presented in Figure 9d and 9e, for 0 µm and 2000 µm, respectively.

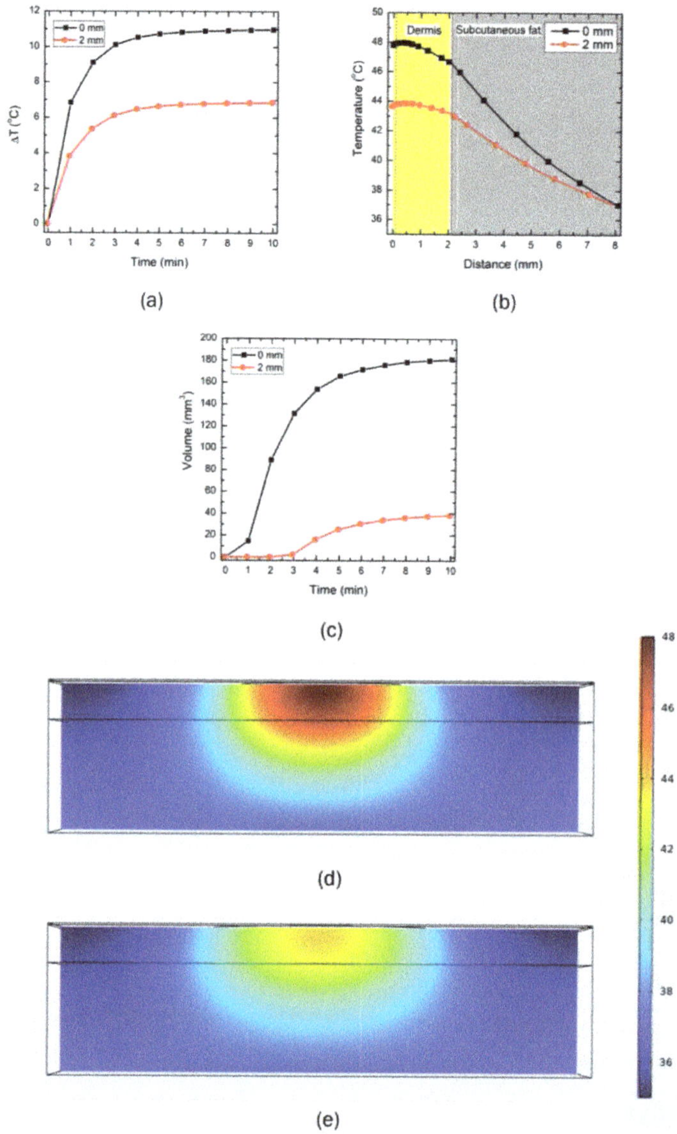

Figure 9. The effect of increasing the distance between the 90° LED (operating at 300 mW) and the skin surface on (**a**) the maximum temperature rise within the skin tissue, (**b**) the temperature distribution along the depth of the skin tissue, and (**c**) the heating volume attained with the T > 43 °C protocol. Temperature distribution (in °C) attained within the skin tissue after 10 min of LED irradiation, when the gap between the LED and skin surface was: (**d**) 0 µm, and (**e**) 2000 µm.

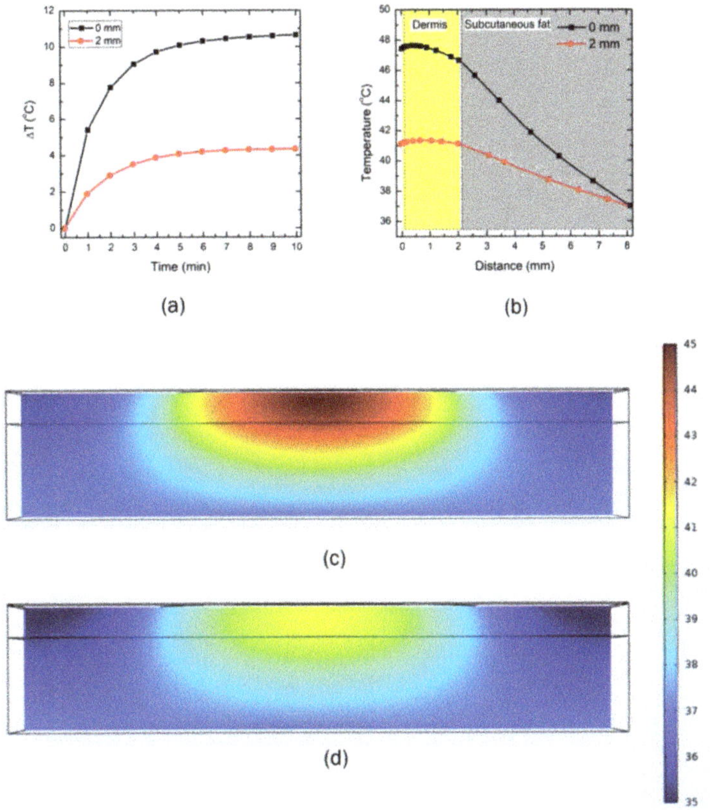

Figure 10. The effect of increasing the distance between the 150° LED (operating at 500 mW) and the skin surface on (**a**) the maximum temperature rise within the skin tissue, and (**b**) the temperature distribution along the depth of the skin tissue. Temperature distribution (in °C) attained within the skin tissue after 10 min of LED irradiation when the gap between the LED and skin surface is: (**c**) 0 μm, and (**d**) 2000 μm.

Similar trends were obtained for the 150° LED, as presented in Figure 10. The maximum temperature dropped by 59% by increasing the distance between the LED and the skin to 2000 μm after 10 min of heating. Moreover, by increasing the distance, the temperature attained at (a) the epidermis-dermis interface dropped by 13.31%, and (b) the dermis-subcutaneous fat interface by 11.8%. Furthermore, the standard deviation of the 150° LED beam Gaussian profile (σ) increased from 7 mm to 10.7 mm as the LED was moved at the distance of 2000 μm from the skin surface, thereby leading to a decline in the irradiation intensity (Equation (3)). The temperature distribution attained within the skin tissue after 10 min of LED therapy is presented in Figure 10c,d. As evident from these figures, as the distance between the LED and skin increased to 2000 μm, the target temperature of 43 °C was not attained. Thus, the distance between the LED and the skin surface significantly affected the efficacy of the proposed LED-based therapy. This highlights that these distances can be altered to control the maximum temperature within the skin tissue, to provide patient-specific desired therapeutic effects.

3.4. Effect of Increase in the Number of LEDs on Temperature Rise within Skin Tissue

The analysis previously presented is made under the assumption that the precise location of the acupoint was previously known. Hence, the goal was to provide the thermal stimulus at the acupoint utilizing near-infrared LED that would further degranulate the highly concentrated mast cells. This would initiate biological mechanisms that would enhance vascular permeability and nerve activation, thereby providing therapeutic effects. The proposed LED-based therapy can also be extended to situations where the acupoint's precise location is unknown. In such a scenario, the treatment goal would be to provide thermal stimulation to the large tissue area in contact with the LED, to achieve therapeutic effects. To attain this, a single LED would be insufficient, and multiple LEDs would have to be used and operated simultaneously, to heat up the large area/volume of the skin tissue. Thus, to include this in our numerical analysis, we present a case study highlighting the enhancement of the heated volume by using multiple LEDs operated simultaneously with the same power. We will proceed to present the results obtained with multiple 90° LEDs driven at the input power of 300 mW for 10 min. Figure 11 presents the temporal variation in the volume obtained using the isotherm of 43 °C, using 1, 2, 3, 4, and 5 LEDs. As evident from Figure 11, the increase in the number of LEDs leads to a substantial increase in the heated volume within the skin tissue. Compared to the single LED, the heated volume after 10 min of LED therapy increased by 324%, 674%, 1014%, and 1338% with the simultaneous usage of 2, 3, 4, and 5 LEDs, respectively. Moreover, the time required to attain the desired temperature of 43 °C also decreased with the increase in the number of LEDs, as shown in Figure 11.

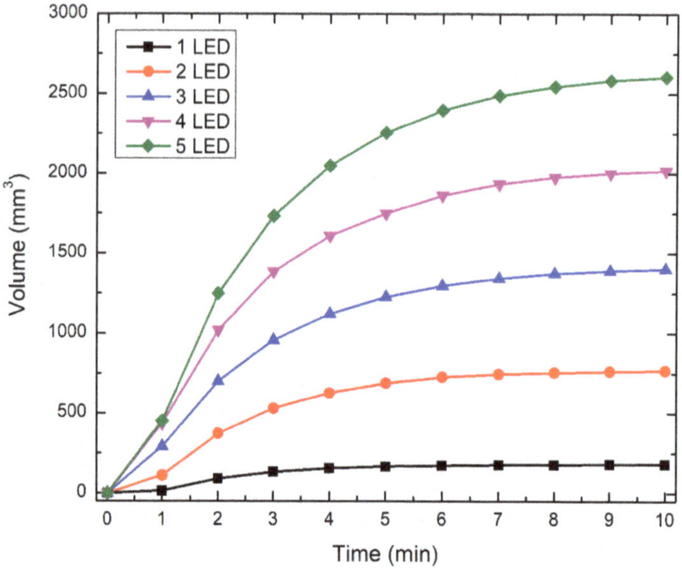

Figure 11. The effect of increased LEDs on the volume attained with the isotherm of 43 °C.

A pictorial representation of the arrangement of these multiple LEDs is presented in Figure 12. Notably, the center-to-center distance of any LED from the one placed at the center was fixed at 10 mm for all configurations. Figure 12 also presents the heated volume as viewed from the top skin surface where the LED is applied, to give a pictorial representation of the volume enhancement with multiple LEDs. This analysis can be further extended to (a) optimize the distance between the LEDs, and (b) operating different LEDs at different powers, so as to restrict the maximum temperature attained within the skin tissue within safe limits, and provide the desired therapeutic effects.

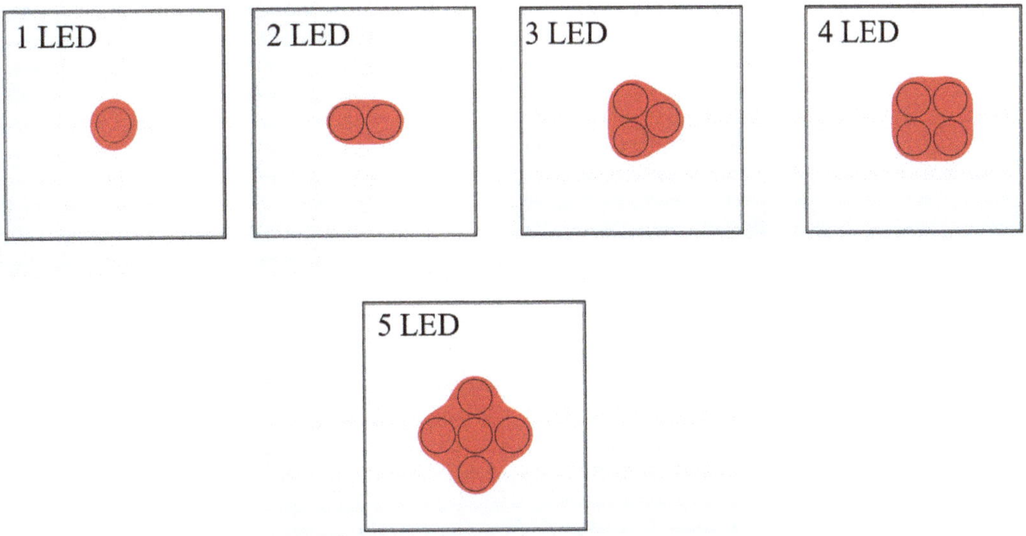

Figure 12. Pictorial representations of the top view of the isotherm of 43 °C on the skin surface attained with different numbers of LEDs.

4. Discussion

In this work, we proposed a novel acupuncture-like therapy, utilizing non-invasive LEDs to provide a physical stimulus to the skin tissue. The proposed therapy is completely non-invasive, eliminating the need to insert needles, as in conventional acupuncture therapy, or the production of smoke and odor from burning moxa sticks, as in conventional moxibustion. The application of visible blue and red LEDs is quite prevalent in several dermatological applications. In this work, we conducted a feasibility study to explore the effectiveness of the near-infrared LED to provide a thermal stimulus to the skin tissue. A computational framework was developed to quantify the spatiotemporal temperature distribution within the multilayer skin tissue irradiated by the near-infrared LED. Two commercially available LEDs with 90° and 150° FWHM angles operating at the wavelength of 850 nm were selected. Firstly, we experimentally characterized the irradiance profile of both these LEDs. A Gaussian profile was obtained for both LEDs, as presented in Figure 2. Secondly, we conducted the experimental tests utilizing both LEDs on the agar phantom gel, to obtain the temperature profile at a depth of 2 mm from the surface. The obtained profile was compared with the results predicted by the developed numerical model, incorporating the thermal and optical properties of the agar phantom under the same environmental conditions. A good agreement was obtained between the numerically predicted and experimentally obtained temperature profiles for both LEDs, as shown in Figure 4. Lastly, once the validation was complete, the numerical model was extended to incorporate multilayer skin tissue, utilizing the well-characterized properties for different layers of skin tissue available in the recent literature. The experimentally characterized Gaussian irradiance profile of the LED was applied at the top surface of the skin tissue to investigate the heat transfer attained within the skin tissue due to light-tissue interactions.

As this is a feasibility study, it is quite important to understand the influence of critical factors that could affect the efficacy of the proposed LED-based photothermal therapy. Thus, parametric studies were conducted to quantify the effects of the LED input power, treatment duration, and LED distance from the surface on the temperature distribution attained within the skin tissue for both the selected LEDs. The results of the parametric analysis of the above-mentioned critical factors are presented in Figures 5–10. These results can be utilized to provide a safe optimal operating condition on the basis of the maximum temperature attained within the skin tissue. It is noteworthy to mention that the goal of the proposed therapy is to provide a thermal stimulus lower than the ablative temperature value of 50 °C, so as to minimize any irreversible damage to the skin tissue. Finally, this study also presented a case study related to the usage of multiple LEDs operating simultaneously, to enhance the volume/area of the thermal stimulus within the skin tissue, as shown in Figures 10 and 11.

The main limitation of the developed model is related to the validation aspect. The developed numerical model was experimentally validated with the results obtained from the agar gel, the optical and thermal properties of which are inconsistent with skin tissue. Additionally, the validation was carried out in ex vivo settings, neglecting the critical impact of microvascular blood perfusion. Our group's current research efforts are significantly devoted to addressing these limitations. Despite these limitations, we still believe that the results reported in this study will assist in better understanding the light-tissue interactions during LED therapy. Moreover, the parametric analysis results have highlighted several critical factors that can be targeted to enhance the efficacy of LED-based heating, to attain the desired therapeutic effects. For example, the linkage between the operating power of the LED and the maximum temperature attained within the skin tissue has been provided. Such information would be critical for designing an LED-based device capable of working under safe temperature limits. The results of the power-temperature relations presented in this study can be further utilized for developing different control algorithms to keep the maximum temperature within the skin tissue at safe operating conditions. These controllers (e.g., a real-time temperature feedback controller or power controller) can be designed by monitoring the temperature at the surface of the skin tissue, and modulating the applied power to keep the maximum temperature below the desired maximum value, to avoid any damage to the skin tissue, thus enhancing both the safety and efficacy of the proposed LED-based therapy. Our group is also working on the clinical translation of the reported numerical results in this work. We are working on developing a miniature device that could be easily mounted onto textile-based wearable products (e.g., a glove, head/arm/wrist/chest band, knee/neck pad, socks), and would be beneficial in providing on-demand physical stimulus to the wearer at home.

5. Conclusions

This study reports the development of a computational framework for accurately quantifying the heat transport within skin tissue exposed to near-infrared LED-based therapy. The numerical validation of the developed model has been carried out by performing experimental tests on agar phantom. A good agreement between the experimental and numerically predicted results was found. The numerical results evaluated the effect of the power, duration, distance between the LED and the skin, and number of LEDs on the maximum temperature attained within the skin tissue. It was found that the operating power and emission angle of the LED significantly affected the spatiotemporal temperature distribution obtained during LED-based therapy. Furthermore, the maximum temperature attained within the skin tissue could be controlled by altering the location between the skin surface and the LED. The application of multiple LEDs would result in an abrupt increase in the heated volume during LED-based photothermal heating, which could be optimized by using different distances between LEDs or operating LEDs at different power. It was found that the proposed LED-based therapy could safely provide thermal stimulus to the skin tissue, to a depth of 4–5 mm. This analysis will pave the way for quantifying the optimum settings for LED therapy, to operate within safe temperature limits. We expect that the future extension and clinical translation of this model will assist in designing a low-cost, miniature device capable of effectively and safely providing a thermal stimulus at home.

Author Contributions: S.S.: conceptualization, methodology, simulations, formal analysis, investigation, data curation, writing—original draft preparation, writing—review and editing; A.E., Z.W. and C.R. contributed to the methodology, experiments and validation; C.-Q.X. and Z.Z. contributed to the conceptualization of the research idea, the supervision of the study, the review and editing of the final manuscript, and funding acquisition. All authors have read and agreed to the published version of the manuscript.

Funding: This research was funded by the National Research Council (NRC) Canada through the Aging in Place program (project number AIP-003).

Institutional Review Board Statement: Not applicable.

Informed Consent Statement: Not applicable.

Data Availability Statement: All data generated or analyzed during this study are included in this article.

Acknowledgments: S.S. and C.-Q.X. would like to acknowledge CMC Microsystems for the provision of products and services that facilitated this research.

Conflicts of Interest: The authors declare no conflict of interest.

References

1. Chow, S.L.; Bozkurt, B.; Baker, W.L.; Bleske, B.E.; Breathett, K.; Fonarow, G.C.; Greenberg, B.; Khazanie, P.; Leclerc, J.; Morris, A.A. Complementary and Alternative Medicines in the Management of Heart Failure: A Scientific Statement From the American Heart Association. *Circulation* **2023**, *147*, e4–e30. [CrossRef]
2. Esmail, N. *Complementary and Alternative Medicine*; Fraser Institute: Vancouver, BC, Canada, 2017.
3. Keene, M.R.; Heslop, I.M.; Sabesan, S.S.; Glass, B.D. Complementary and alternative medicine use in cancer: A systematic review. *Complement. Ther. Clin. Pract.* **2019**, *35*, 33–47. [CrossRef] [PubMed]
4. Milenkovic, J. Complementary and alternative medicine in European countries-legislative framework. *Tradit. Med. Res.* **2020**, *5*, 125. [CrossRef]
5. Ng, J.Y.; Bhatt, H.A.; Raja, M. Complementary and alternative medicine mention and recommendations in pancreatic cancer clinical practice guidelines: A systematic review and quality assessment. *Integr. Med. Res.* **2023**, *12*, 100921. [CrossRef]
6. Sunjka, M.L.; Pejcic, A.; Jakovljevic, M. Utilization patterns of complementary and alternative medicine in Australia, Canada and the United States: Popularity of dietary supplements, mind-body and manipulative therapies. *Farmeconomia Health Econ. Ther. Pathw.* **2017**, *18*, 15–20.
7. Wode, K.; Henriksson, R.; Sharp, L.; Stoltenberg, A.; Hök Nordberg, J. Cancer patients' use of complementary and alternative medicine in Sweden: A cross-sectional study. *BMC Complement. Altern. Med.* **2019**, *19*, 62. [CrossRef] [PubMed]
8. Austin, S.; Ramamonjiarivelo, Z.; Qu, H.; Ellis-Griffith, G. Acupuncture use in the United States: Who, where, why, and at what price? *Health Mark. Q.* **2015**, *32*, 113–128. [CrossRef] [PubMed]
9. Lu, L.; Zhang, Y.; Tang, X.; Ge, S.; Wen, H.; Zeng, J.; Wang, L.; Zeng, Z.; Rada, G.; Ávila, C. Evidence on acupuncture therapies is underused in clinical practice and health policy. *BMJ* **2022**, *376*, e067475. [CrossRef]
10. Xia, Y.; Ding, G.; Wu, G.-C. *Current Research in Acupuncture*; Springer: Berlin/Heidelberg, Germany, 2013.
11. Jun, M.-H.; Kim, Y.-M.; Kim, J.U. Modern acupuncture-like stimulation methods: A literature review. *Integr. Med. Res.* **2015**, *4*, 195–219. [CrossRef]
12. Longhurst, J.C. Defining meridians: A modern basis of understanding. *J. Acupunct. Meridian Stud.* **2010**, *3*, 67–74. [CrossRef]
13. Ma, Y.; Dong, M.; Zhou, K.; Mita, C.; Liu, J.; Wayne, P.M. Publication trends in acupuncture research: A 20-year bibliometric analysis based on PubMed. *PLoS ONE* **2016**, *11*, e0168123. [CrossRef] [PubMed]
14. Li, Y.; Yu, Y.; Liu, Y.; Yao, W. Mast cells and acupuncture analgesia. *Cells* **2022**, *11*, 860. [CrossRef] [PubMed]
15. Gao, X.; Wang, Y.; Meng, H.; Li, S.; Jiang, H.; Zhang, Z.; He, J.; Zhao, Y.; Zhang, S.; Zhai, W. Acupuncture for brain diseases: Conception, application, and exploration. *Anat. Rec.* 2022. [CrossRef] [PubMed]
16. Kim, K.-W.; Hong, S.; Kim, H.S.; Kim, T.; Ahn, J.; Song, H.-S.; Kim, Y.-K.; Oh, J.-Y.; Hwang, T.-Y.; Lee, H. Physiological impact of nanoporous acupuncture needles: Laser Doppler perfusion imaging in healthy volunteers. *PLoS ONE* **2019**, *14*, e0226304. [CrossRef]
17. Zhu, J.; Li, J.; Yang, L.; Liu, S. Acupuncture, from the ancient to the current. *Anat. Rec.* **2021**, *304*, 2365–2371. [CrossRef]
18. Yang, W.; Liu, X.; Zhang, X.; Li, C.; Li, Z.; Li, Y.; Li, M. Bibliometric analysis of acupuncture and moxibustion treatment for mild cognitive impairment. *Front. Neurosci.* **2023**, *17*, 1209262. [CrossRef]
19. Zhang, Z.-J.; Wang, X.-M.; McAlonan, G.M. Neural acupuncture unit: A new concept for interpreting effects and mechanisms of acupuncture. *Evid. Based Complement. Altern. Med.* **2012**, *2012*, 429412. [CrossRef]
20. Deng, H.; Shen, X. The mechanism of moxibustion: Ancient theory and modern research. *Evid. Based Complement. Altern. Med.* **2013**, *2013*, 379291. [CrossRef]
21. Liu, H.; Huang, Z.; Wei, L.; Huang, H.; Li, Q.; Peng, H.; Liu, M. Uncertainty analysis and optimization for mild moxibustion. *PLoS ONE* **2023**, *18*, e0282355. [CrossRef]

22. Li, Y.; Sun, C.; Kuang, J.; Ji, C.; Feng, S.; Wu, J.; You, H. An in vitro and numerical study of moxibustion therapy on biological tissue. *IEEE Trans. Biomed. Eng.* **2017**, *65*, 779–788. [CrossRef]
23. Sun, C.; Ji, C.; Li, Y.; Kuang, J.; Wu, J. A comparison study of photothermal effect between moxibustion therapy and laser irradiation on biological tissue. *Int. J. Therm. Sci.* **2021**, *164*, 106924. [CrossRef]
24. Liu, H.; Huang, Z.; Wei, L.; Huang, H.; Li, Q.; Peng, H.; Liu, M. Numerical Simulation of Temperature Distribution during Mild Moxibustion. *Evid. Based Complement. Altern. Med.* **2022**, *2022*, 6604372. [CrossRef] [PubMed]
25. Kim, G.; Hwang, Y.-I.; Ryu, Y.; Kim, H.-J.; Bae, Y.-M.; Kim, K.-B. Ultrasonic device developed for non-invasive moxibustion therapy. *Integr. Med. Res.* **2021**, *10*, 100729. [CrossRef]
26. Solovchuk, M.; Deng, H.-A.; Sheu, T.W. Experimental and numerical study on the temperature elevation in tissue during moxibustion therapy. *Evid. Based Complement. Altern. Med.* **2020**, *2020*, 7514302. [CrossRef]
27. Sun, C.; Li, Y.; Kuang, J.; Liang, X.; Wu, J.; Ji, C. The thermal performance of biological tissue under moxibustion therapy. *J. Therm. Biol.* **2019**, *83*, 103–111. [CrossRef]
28. Tsuruoka, N.; Watanabe, M.; Seki, T.; Matsunaga, T.; Hagaa, Y. Acupoint stimulation device using focused ultrasound. In Proceedings of the 2010 Annual International Conference of the IEEE Engineering in Medicine and Biology, Buenos Aires, Argentina, 31 August–4 September 2010; pp. 1258–1261.
29. Tsuruoka, N.; Watanabe, M.; Takayama, S.; Seki, T.; Matsunaga, T.; Haga, Y. Brief effect of acupoint stimulation using focused ultrasound. *J. Altern. Complement. Med.* **2013**, *19*, 416–419. [CrossRef] [PubMed]
30. Calderhead, R.G.; Vasily, D.B. Low level light therapy with light-emitting diodes for the aging face. *Clin. Plast. Surg.* **2016**, *43*, 541–550. [CrossRef] [PubMed]
31. Rossi, F.; Pini, R.; De Siena, G.; Massi, D.; Pavone, F.S.; Alfieri, D.; Cannarozzo, G. A blue-LED-based device for selective photocoagulation of superficial abrasions: Theoretical modeling and in vivo validation. In *Photonic Therapeutics and Diagnostics VI: 23–25 January 2010, San Francisco, CA, USA*; SPIE: Bellingham, DC, USA, 2010; pp. 53–59.
32. Sorbellini, E.; Rucco, M.; Rinaldi, F. Photodynamic and photobiological effects of light-emitting diode (LED) therapy in dermatological disease: An update. *Lasers Med. Sci.* **2018**, *33*, 1431–1439. [CrossRef]
33. Nasouri, B.; Murphy, T.E.; Berberoglu, H. Simulation of laser propagation through a three-layer human skin model in the spectral range from 1000 to 1900 nm. *J. Biomed. Opt.* **2014**, *19*, 075003. [CrossRef]
34. Dremin, V.; Novikova, I.; Rafailov, E. Simulation of thermal field distribution in biological tissue and cell culture media irradiated with infrared wavelengths. *Opt. Express* **2022**, *30*, 23078–23089. [CrossRef]
35. Wongchadakul, P.; Rattanadecho, P.; Wessapan, T. Implementation of a thermomechanical model to simulate laser heating in shrinkage tissue (effects of wavelength, laser irradiation intensity, and irradiation beam area). *Int. J. Therm. Sci.* **2018**, *134*, 321–336. [CrossRef]
36. Andreozzi, A.; Brunese, L.; Iasiello, M.; Tucci, C.; Vanoli, G.P. Modeling heat transfer in tumors: A review of thermal therapies. *Ann. Biomed. Eng.* **2019**, *47*, 676–693. [CrossRef] [PubMed]
37. Bhowmik, A.; Singh, R.; Repaka, R.; Mishra, S.C. Conventional and newly developed bioheat transport models in vascularized tissues: A review. *J. Therm. Biol.* **2013**, *38*, 107–125. [CrossRef]
38. Singh, S.; Melnik, R. Thermal ablation of biological tissues in disease treatment: A review of computational models and future directions. *Electromagn. Biol. Med.* **2020**, *39*, 49–88. [CrossRef]
39. Hasgall, P.; Di Gennaro, F.; Baumgartner, C.; Neufeld, E.; Lloyd, B.; Gosselin, M.; Payne, D.; Klingenböck, A.; Kuster, N. IT'IS Database for Thermal and Electromagnetic Parameters of Biological Tissues. Available online: itis.swiss/database (accessed on 7 March 2023).
40. Mohammadi, A.; Bianchi, L.; Korganbayev, S.; De Landro, M.; Saccomandi, P. Thermomechanical modeling of laser ablation therapy of tumors: Sensitivity analysis and optimization of influential variables. *IEEE Trans. Biomed. Eng.* **2021**, *69*, 302–313. [CrossRef]
41. Salomatina, E.; Jiang, B.; Novak, J.; Yaroslavsky, A.N. Optical properties of normal and cancerous human skin in the visible and near-infrared spectral range. *J. Biomed. Opt.* **2006**, *11*, 064026. [CrossRef] [PubMed]
42. Proskurnin, M.A.; Khabibullin, V.R.; Usoltseva, L.O.; Vyrko, E.; Mikheev, I.V.; Volkov, D.S. Photothermal and optoacoustic spectroscopy: State of the art and prospects. *Phys. Uspekhi* **2022**, *65*, 270. [CrossRef]
43. Paul, A.; Narasimhan, A.; Kahlen, F.J.; Das, S.K. Temperature evolution in tissues embedded with large blood vessels during photo-thermal heating. *J. Therm. Biol.* **2014**, *41*, 77–87. [CrossRef] [PubMed]
44. Patapoutian, A.; Peier, A.M.; Story, G.M.; Viswanath, V. ThermoTRP channels and beyond: Mechanisms of temperature sensation. *Nat. Rev. Neurosci.* **2003**, *4*, 529–539. [CrossRef] [PubMed]
45. Adriaensen, H.; Gybels, J.; Handwerker, H.; Van Hees, J. Response properties of thin myelinated (A-delta) fibers in human skin nerves. *J. Neurophysiol.* **1983**, *49*, 111–122. [CrossRef]
46. Wang, G.-Y.; Wang, L.-L.; Xu, B.; Zhang, J.-B.; Jiang, J.-F. Effects of moxibustion temperature on blood cholesterol level in a mice model of acute hyperlipidemia: Role of TRPV1. *Evid. Based Complement. Altern. Med.* **2013**, *2013*, 871704. [CrossRef] [PubMed]

bioengineering

MDPI

Communication

Spatial Dependence of Log-Transformed Electromyography–Force Relation: Model-Based Sensitivity Analysis and Experimental Study of Biceps Brachii

Chengjun Huang, Maoqi Chen, Zhiyuan Lu, Cliff S. Klein and Ping Zhou

[1] Department of Neuroscience, Baylor College of Medicine, Houston, TX 77030, USA
[2] School of Rehabilitation Science and Engineering, University of Health and Rehabilitation Sciences, Qingdao 266072, China
[3] Guangdong Work Injury Rehabilitation Center, Rehabilitation Research Institute, Guangzhou 510440, China
* Correspondence: dr.ping.zhou@outlook.com

Abstract: This study investigated electromyography (EMG)–force relations using both simulated and experimental approaches. A motor neuron pool model was first implemented to simulate EMG–force signals, focusing on three different conditions that test the effects of small or large motor units located more or less superficially in the muscle. It was found that the patterns of the EMG–force relations varied significantly across the simulated conditions, quantified by the slope (b) of the log-transformed EMG-force relation. b was significantly higher for large motor units, which were preferentially located superficially rather than for random depth or deep depth conditions ($p < 0.001$). The log-transformed EMG–force relations in the biceps brachii muscles of nine healthy subjects were examined using a high-density surface EMG. The slope (b) distribution of the relation across the electrode array showed a spatial dependence; b in the proximal region was significantly larger than the distal region, whereas b was not different between the lateral and medial regions. The findings of this study provide evidence that the log-transformed EMG–force relations are sensitive to different motor unit spatial distributions. The slope (b) of this relation may prove to be a useful adjunct measure in the investigation of muscle or motor unit changes associated with disease, injury, or aging.

Keywords: surface electromyography (EMG); motor unit; spatial distribution; EMG-force relation; biceps brachii

Citation: Chengjun Huang, Maoqi Chen, Zhiyuan Lu, Cliff S. Klein and Ping Zhou Spatial Dependence of Log-Transformed Electromyography–Force Relation: Model-Based Sensitivity Analysis and Experimental Study of Biceps Brachii. *Bioengineering* **2023**, *10*, 469. https://doi.org/10.3390/bioengineering10040469

Academic Editors: Sundeep Singh, Roderick Melnik, Esther Pueyo and Aurélien Courvoisier

Received: 17 February 2023
Revised: 20 March 2023
Accepted: 26 March 2023
Published: 12 April 2023

1. Introduction

Muscle force is produced by the contraction of muscle fibers belonging to different motor units. Large motor units have higher activation thresholds with faster and larger twitch (Type II) muscle fibers, compared to small motor units that have lower activation thresholds with slower and smaller twitch (Type I) muscle fibers [1,2]. The distribution of large and small motor units within a muscle has been explored in various studies. Early histochemical studies revealed no significant difference in the proportion of slow and fast muscle fibers between deep and superficial regions of human upper and lower limb muscles [3,4]. However, others reported that fiber type distribution was not random [5],for example, with a predominance of Type II fibers at the muscle surface and Type I fibers in the deeper regions [6–8].

The electrical manifestation of a muscle contraction can be measured by electromyography (EMG), which has an amplitude related, in part, to the location of the motor units relative to the recording electrodes, i.e., the spatial distribution of the motor units. Beck et al. [9] observed changes in the vastus medialis EMG–force relation at various distances from the innervation zone. Trent et al. [10] also found that the shape of the biceps brachii EMG–force relation was altered as a function of the distance between the innervation zone and the recording position. The spatial sensitivity of the EMG–force relation relative to the innervation zone was also confirmed in a linear electrode array study of paretic and contralateral muscles following stroke [11]. Given that the same force signal was used for different channels, the spatial variation in EMG–force relation reflected mainly changes in EMG properties because of changes in the relative position between active motor units and the recording electrodes. While most previous simulation studies have investigated various motor unit factors that may impact the EMG–force relation [12–15], few have examined the effect of motor unit distribution within a muscle. Compared with the often-used random distribution of different motor units or muscle fiber types, Robertson and Johnston applied a weighting term to produce different fiber type content in deep and superficial muscle regions, resulting in a larger variety of EMG–force relations [16].

In this study, quantitative analyses of both simulated and experimental EMG–force relations were performed. First, we hypothesize that the EMG–force relation is sensitive to different motor unit spatial distributions. To test this hypothesis, we implemented a classic motor neuron pool model to simulate the EMG–force relation. Three different motor unit distributions with respect to muscle depth were simulated (superficial, random, and deep). For each condition, a log-transformed EMG–force relation was constructed, which has been proved useful in characterizing EMG or mechanomyographic outputs with respect to the muscle contraction level [10,17–19]. Second, we hypothesize that a muscle's EMG–force relation is sensitive to different electrode positions. To test this hypothesis, we recorded high-density surface EMG signals using electrode arrays at different voluntary isometric contraction levels of the biceps brachii in healthy subjects and examined the log-transformed EMG–force relation across the electrode array recordings. The findings confirmed both hypotheses, i.e., the slope of the log-transformed EMG–force relation was found to be sensitive to different motor unit spatial distributions in the simulation study and to different proximal–distal electrode positions in the experimental study. Therefore, it may provide a convenient and useful adjunct parameter for assessing muscle or motor unit changes associated with disease, injury, or advanced aging.

2. Materials and Methods

2.1. Simulation Study

Surface EMG and force signals were simulated using a motor unit pool model [12]. There were 150 motor units in the modeled motor neuron pool, which innervated about 70,000 muscle fibers. The number of muscle fibers per motor unit and motor unit recruitment threshold (RTE) each had an exponential distribution. Consequently, most motor units had a relatively low RTE and number of muscle fibers (range 22–2137) or twitch force (range 1–100). Once the modeled excitatory drive exceeded RTE, the motor unit started to discharge at a minimum firing rate of 8 Hz [12]. The last motor unit was simulated to be recruited at 80% or 40% maximum excitation. The former corresponded to the activity of large muscles, where recruitment plays a significant role throughout the force range (i.e., biceps brachii), whereas the latter mimicked the activity of small muscles where rate coding plays a more important role in force generation (i.e., hand muscles). The firing rate of each motor unit increased linearly with excitation until the peak firing rate (PFR) was reached. Both "onion skin" and "reverse onion skin" firing strategies were used when simulating motor unit firing behavior. The "onion skin" strategy indicates that early recruited low-threshold motor units attain higher peak firing rates than later-recruited high-threshold motor units. Although the "onion skin" pattern is the most common observation during voluntary muscle contractions [20,21], other patterns of motor unit firing rate behavior have been proposed [22,23]. Some reported similar PFR in all motor units [24] or that high-threshold units reach higher PFR than low-threshold units (reverse "onion skin" pattern) [25,26]. Therefore, both strategies were implemented in our study. The PFR of the largest and smallest motor unit was set to 35 Hz and 25 Hz, respectively. The excitation level ranged from 10% to 100% of maximum excitation with 10% increments.

A tripole model was used to simulate the generation of EMG signals [27]. In brief, the muscle was assumed to have a cylindrical shape. The x-axis represents the width of the muscle in the transverse direction, the *z*-axis represents the muscle fiber direction, and the *y*-axis represents the muscle depth (Figure 1). Muscle diameter was set to 24 mm, and the thickness of fat plus skin layers was 3 mm. The action potential φ_j, generated by fiber *j* and detected by the electrode on the surface (*x*–*z* plane), was given by Equation (1), where K_a, represents the ratio of electric conductivities in the *z* (σ_z) and *x* directions(σ_x) (which were set to $\sigma_z = 0.33$ and $\sigma_x = 0.06$ S/m, respectively) [27]. A motor unit action potential (MUAP) was simulated as the sum of its constituent muscle fiber action potentials. The interspike intervals (ISI) for each motor unit were simulated as a random process with a Gaussian probability function. Surface EMG signals under different excitation levels detected by the electrode were simulated as a sparse combination of MUAP trains from all active motor units.

$$\varphi_j = \frac{1}{2\pi\sigma_x}\sum_{i=1}^{6}\frac{P_i}{\sqrt{((x-x_i)^2+y_i^2)K_a+(z-z_i)^2}} \quad (1)$$

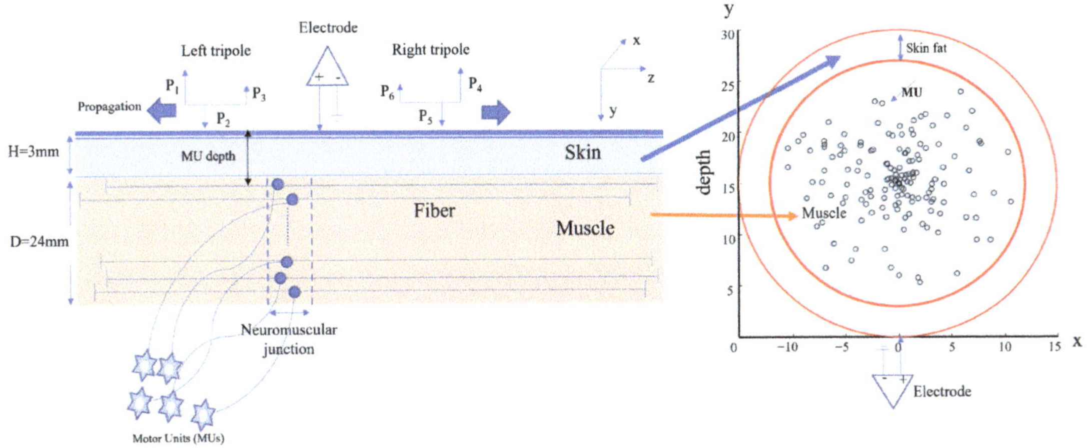

Figure 1. The cylindrical muscle model and electrode for surface EMG simulation. Left panel: the tripole model and the position of the electrode with respect to the cylinder parameters. Right panel: the cross-section of the cylindrical model showing muscle and skin fat layers.

In this study, the electrode was simulated at the midline of the muscle ($x = 0$) (Figure 1) [27]. The sampling rate was 2000 Hz. For each motor unit (the *i*th motor unit), the twitch force simulation was adopted from the Fuglevand model [12], which followed a second-order critically damped impulse response (Equation (2)). The range of peak twitch force P_i across all motor units was 100-fold, and the range of contraction time T_i was 3-fold. Motor units with lower RTE were simulated to have lower and longer twitch forces. For example, the first recruited smallest motor unit had a P_i of 1 arbitrary unit (au) and a T_i of 90 ms, while the largest motor unit had a P_i of 100 au and T_i of 30 ms.

$$f_{i,\,j}(t) = g_{i,j} * \frac{P_i \cdot t}{T_i} \cdot e^{1-\frac{t}{T_i}} \quad (2)$$

The gain $g_{i,j}$ (Equation (2)) was adjusted based on T_i and the ISI of the discharge *j*. The gain was set to 1 when $T_i/ISI_j < 0.4$ and otherwise was determined as in Equation (3). The total muscle contraction force was calculated as the linear summation of each motor unit force output.

$$g_{i,j} = \frac{1-e^{-2(T_i/ISI_j)^3}}{T_i/ISI_j} \qquad \text{when} \quad T_i/ISI_j > 0.4 \quad (3)$$

Three different motor unit distributions, with respect to muscle depth, were simulated. The first of these were larger higher-threshold motor units, which were assumed to be preferentially located in the superficial regions of the muscle (termed "superficial"). In this case, there is a negative correlation between motor unit size and motor unit depth (Figure 2 left panel, slope of the linear regression = −0.13). The second simulation was small and large motor units, which were assumed to be randomly distributed in the muscle (termed "random") regardless of the depth (Figure 2 middle panel, slope of the linear regression close to 0). The third simulation was larger higher-threshold motor units, which were assumed to be preferentially located in the deeper regions of the muscle (termed "deep"). In this case, there was a positive correlation between motor unit size and motor unit depth (Figure 2 right panel, slope of the linear regression = 0.12).

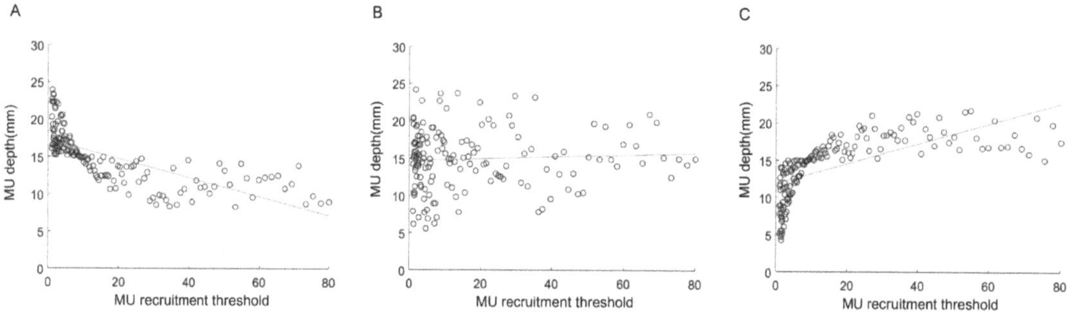

Figure 2. Three different motor unit distributions with respect to muscle depth. (**A**) motor unit depth is negatively related with motor unit size. (**B**) motor unit depth is randomly assigned. (**C**) motor unit depth is positively related to motor unit size.

2.2. *Experimental Study*

2.2.1. Participants and Consent

Nine subjects (8 male, 1 female, 28.9 ± 4.8 years, mean ± SD) participated in the experiment. They were recruited by word of mouth or posters and had no known history or symptoms of any neuromuscular disorder. All subjects were well informed of the experimental procedures, risks, and possible discomfort and provided written informed consent. The study was approved by the local ethics committee (ethical approval number: GWIRC-AF/SC-07/2016.20).

2.2.2. Experiment Protocol

Surface EMG signals were recorded from the biceps brachii muscle using two electrode arrays (Figure 3A,B). Each array (ELSCH064NM2, Bioelettronica, Torino, Italy) consists of 64 channels with an 8 mm inter-channel distance arranged in a grid of 5 columns by 13 rows (one column consisted of 12 channels and the other four of 13 channels). One array was placed over the lateral side (Matrix 1) and another over the medial side (Matrix 2) of the biceps brachii and secured with elastic straps. The columns of both arrays were placed parallel to the fiber direction. A ground channel was placed at the elbow. Subjects sat in height-adjustable chairs, each with the back fully against the backrest (Figure 3C). The dominant forearm (right side in all cases) was placed in a force-measuring device (CZL-3 T, Leitai, Bengbu, China) used to measure elbow flexion force. The elbow angle was set at 120° degrees (180° = full extension), and the forearm was supinated. The generated force was displayed on a second monitor in front of the subject as real-time feedback. Each subject performed two to three maximum voluntary contraction (MVC) trials, and the largest value of the trials was adopted as the MVC value. During the contractions, subjects were strongly encouraged to give their best effort without altering the position of the arm significantly. Each subject completed a series of 5 s submaximal contractions at 20, 40, 60, and 80% MVC, with a brief rest period (>3 min) between each to avoid muscle fatigue.

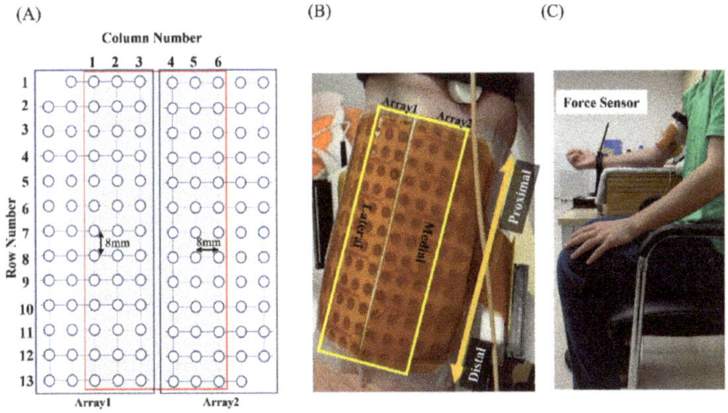

Figure 3. (**A**) Schematic representation of the two adhesive 2D matrices for experimental signals. (**B**) High-density electrode matrix consisting of a grid with 6 columns (that are positioned parallel to the muscle fiber direction) and 13 rows. (**C**) Illustration of the experimental setup.

Surface EMG signals were recorded by a signal amplifier in a monopolar configuration (EMG–USB2, with a sampling frequency of 2048 Hz, 12-bit A/D converter, Bioelettronica, Torino, Italy). The EMG amplitude and mean force were calculated from a 2 s segment (4000 sample points) of each contraction. The force and EMG amplitude were normalized to their maximum levels.

2.3. Log-Transformed EMG–Force Relation

When we have force and surface EMG measurement of a muscle's isometric contraction at different levels, a linear regression model fitting can be used to describe the natural log-transformed EMG–force relation, as shown in Equation (4):

$$\ln[Y] = b(\ln[X]) + \ln[a] \tag{4}$$

where X is the force measurement, Y is the root mean square amplitude (RMS) of surface EMG, ln represents the natural log, b is the slope, and $\ln[a]$ is the natural log of the Y intercept. The equation can also be expressed as an exponential equation after antilog transformation:

$$Y = aX^b \tag{5}$$

The value of b indicates whether the original, non-transformed EMG–force relation is linear or nonlinear. If b is equal to 1, then Y increases linearly with X, if b is greater than 1, Y increases faster than X, and if b is less than 1, Y increases slower than X.

2.4. Data Analysis

For the simulation analysis, 50 repetitions were performed and averaged for each depth condition. A one-way ANOVA was performed to determine the effect of the depth condition on the exponent b (slope of the EMG–force relation in Equation (4)). The distribution of data was tested using the Kolmogorov–Smirnov normality test. When a significant overall effect was confirmed, the Bonferroni-corrected post-hoc test for multiple comparisons was performed.

In the experimental study, for each subject, we averaged b across the whole electrode array and the different areas of the array as well. The proximal area of the muscle was represented by electrodes in rows 3 and 4, and the distal area was represented by rows 10 and 11. The lateral area of the muscle was represented by electrode columns 1 and 2, and the medial area by columns 5 and 6. The Wilcoxon signed-rank test was used to test whether b was significantly different between the proximal and distal regions or between the lateral and medial areas. In all of the statistical analyses, a p-value less than 0.05 was considered to be statistically significant. The analyses were performed using MATLAB 2020b.

3. Results

3.1. Simulation of the EMG–Force Relation

Different motor unit recruitment ranges (40% and 80%of maximum excitation) and firing strategies ("onion skin" and reverse "onion skin") were used in the motor neuron pool model. Although variation in these factors influenced the simulated EMG–force relation, as demonstrated in previous simulation studies [12–15], our data confirmed a similar trend of the slope changes of the log-transformed EMG–force relation with respect to different motor unit depth distributions (using the simulated two motor unit recruitment ranges and two firing strategies). For the sake of simplicity, the results from only one motor unit recruitment range (40% of maximum excitation) and firing strategy (reverse "onion skin") are presented below.

Representative examples of the log-transformed EMG–force relation under the three simulated motor unit depth distributions are shown in Figure 4. Motor unit depth distribution was found to have a significant main effect on b ($p < 0.001$); b was largest when larger motor units were located in the more superficial regions of the muscle (and the smaller units were located in the deeper regions). Post-hoc comparisons indicate that b for large motor units preferentially located superficially (1.38 $\pm$ 0.03, mean $\pm$ SD) was significantly higher than for the random depth (1.00 $\pm$ 0.07, $p < 0.001$) or deep depth (0.94 $\pm$ 0.02, ($p < 0.001$) conditions. The b for the random depth condition was also significantly larger than the deep depth condition ($p < 0.001$).

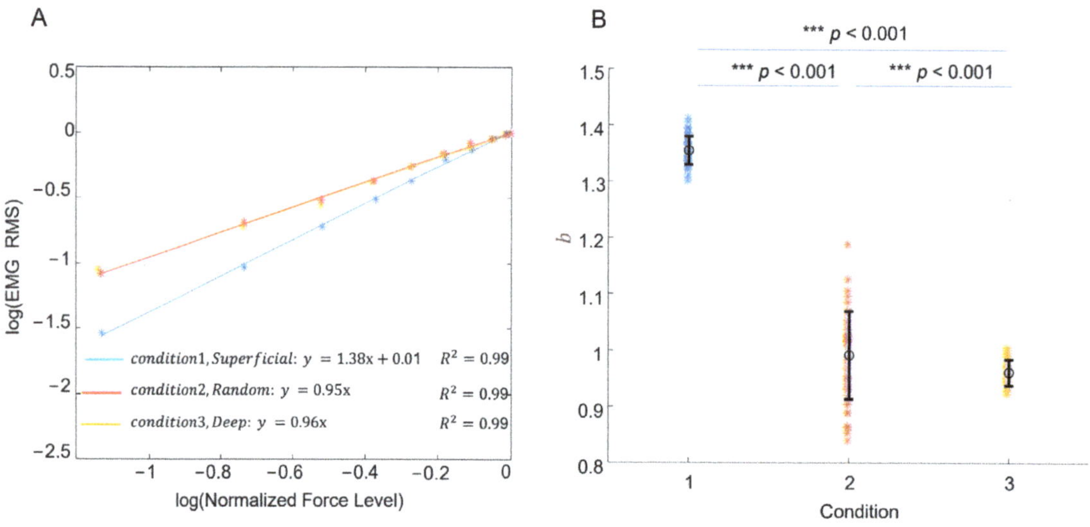

Figure 4. (**A**) Representative examples of log transformed EMG RMS–force relation when large motor units were assumed to preferentially located in the superficial or deep regions of the muscle or unrelated to muscle depth (random). (**B**) The results for b of all the repetitions in each simulated condition.

3.2. Experimental Recording of the EMG–Force Relation

The log-transformed normalized EMG–force relation from one representative subject is presented in Figure 5. EMG signals were averaged from all electrodes to represent overall muscle activity; b was 1.21, indicating that EMG amplitude increased relatively more than force (Figure 5A). The distribution of b across all channels is shown in Figure 5B for this subject. In this instance, b in the proximal region tended to be larger than in the distal region, whereas b was similar in the lateral and medial regions. In all subjects, b was larger in the proximal than distal region; mean *b* was larger proximally than distally 1.14 ± 0.11 and 1.04 ± 0.14, Z = 2.67, $p < 0.01$, Figure 6A). The effect size was 0.89, calculated as Z statistic divided by the square root of the sample size. No significant difference in *b* was found between the lateral (1.08 ± 0.13) and medial (1.09 ± 0.12) regions (Figure 6B).

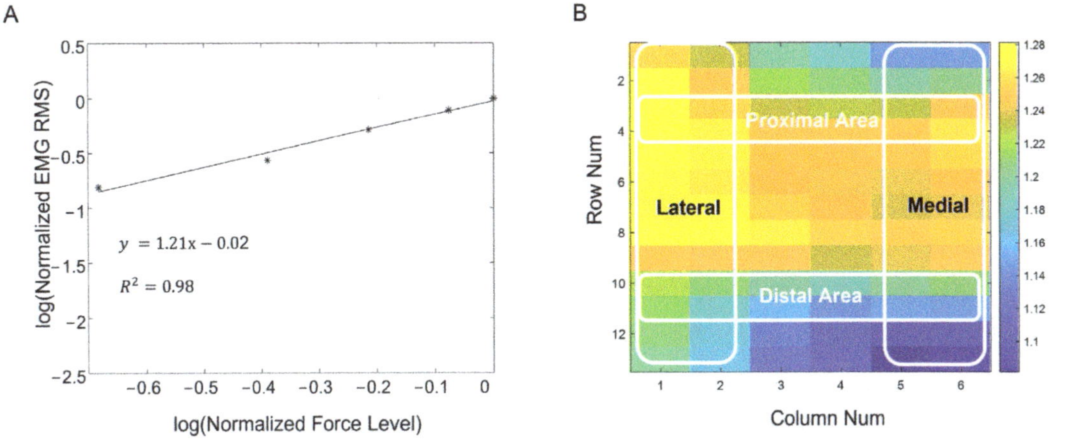

Figure 5. (**A**) The log transformed normalized EMG RMS–force relation (averaged from all electrodes) in a representative subject. (**B**) The distribution of *b* across all electrodes for the subject.

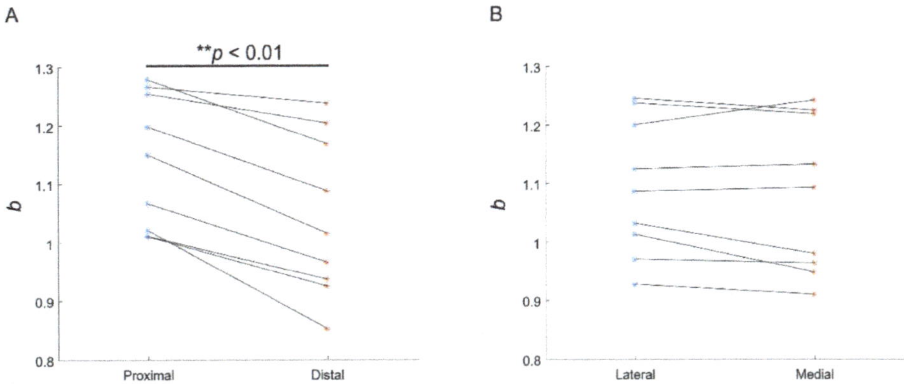

Figure 6. (**A**) Individual *b* for all subjects recorded from the proximal (average of all channels in rows 3 and 4) and distal (rows 10 and 11) muscle regions. (**B**) Individual *b* for all subjects recorded from the lateral (average of all channels in columns 1 and 2) and medial (columns 5 and 6) muscle regions.

4. Discussion

This study examined the EMG–force relation with respect to the depth distributions of different motor units in a muscle. The relation between EMG amplitude and muscle force has been extensively studied by others [12–16]. Two typical forms of the EMG–force relation have been reported [28], namely an approximately linear relation and a curvilinear relation, in which EMG amplitude increases relatively faster than force. In this study, the log transformation was used to process the EMG–force relation. The resultant slope (*b* term) can be used to characterize both linear and nonlinear forms of the relationship (or linear, logarithmic, and exponential forms [16]). Log-transformed processing has been used in previous studies when examining EMG–force relations or the mechanomyographic amplitude versus force relations [17–19].

We found that simulated b of the transformed EMG–force relation was sensitive to different motor unit depth distributions. The predominance of large motor units in the more superficial regions of the muscle tended to increase the slope. The muscle EMG–force relation reflected progressive changes in the recruitment and firing rates of the constituent motor units. The variation in b, due to different motor unit depth distributions, arose from EMG changes rather than changes in force. During a muscle contraction, small motor units were first recruited at relatively low forces, and large motor units were progressively activated as the force was increased. The volume conductor effect of skin and subcutaneous tissues impacted upon EMG amplitude and b may have been influenced by motor unit depth distribution, i.e., greater b in muscles with relatively larger superficial motor units. Therefore, EMG amplitude can be relatively small during low forces because less of the small (deep) motor unit potentials can be volume conducted to the muscle surface. As force is progressively increased, EMG amplitude can increase relatively more than force because more of the large (superficial) motor unit potentials can be volume conducted to the muscle surface.

The spatial distribution of muscle fiber types has been studied previously, and the findings are mixed. Some reported that Type I and Type II fibers were randomly distributed in the muscle cross-section [3,4]. Others found a larger proportion of Type II fibers at the muscle surface [6–8]. In the biceps brachii, Elder and colleagues reported a higher percentage of Type II fibers in the superficial than in the deep region in all four people examined at an autopsy (group means, 58.1% vs. 52.3%), but the difference was not significant based on an ANOVA analysis [5]. However, the difference was significant when based on a paired t-test ($p = 0.01$). Similarly, in another study of six people at an autopsy, the mean biceps brachii Type II fiber percentage was higher in the superficial than in the deep region (57.7% vs. 49.5%), but statistics were not reported for these means [6]. In macro-EMG(which captures the activity of a complete motor unit) analysis at different depths of the human vastus lateralis muscle, large motor units were found to be more superficially distributed, indicative of an inverse relationship between recording depth and macro-EMG amplitude or area [29,30]. In contrast, there are also EMG studies suggesting that large motor units are located deeper in a muscle [31–33].

The high-density surface EMG used in this study provided an opportunity to observe the slope distribution of the log-transformed EMG–force relation across the electrode array. Our experimental results indicate that b was larger proximally than distally, whereas it was not different between the medial and lateral areas of the muscle. This is consistent with a previous linear electrode array surface EMG examination of leg extensors, where b was significantly higher near innervation zone channels than distal channels [10]. Although a depth gradient with motor unit size has been suggested, few studies have examined how such a gradient might change in different muscle regions. Based on our simulation results, the relatively high b of the log-transformed EMG–force relation observed from the proximal biceps seems to suggest a difference in motor unit spatial distribution between proximal and distal regions. From the experimental results, we might infer that muscle fibers may not be parallel to the skin over the whole muscle (fiber) length; however, they may run at an angle. Thus, fast-twitch fibers might be located more superficially in the proximal region and extend deeper into the distal portion of the muscle. Alternatively, the apparent proximal–distal dependency of b may simply reflect gross anatomy and the location of proximal–distal muscle regions relative to the recording electrodes.

Experimental studies of the EMG–force relation may have value in understanding motor unit plasticity associated with aging, disease, and injury. For example, changes in EMG–force relation have been reported in stroke [34–37], amyotrophic lateral sclerosis [38], and spinal cord injury [39] patients. A number of motor unit factors may collectively influence the EMG–force relation, including loss of functional motor units, impaired motor unit control properties, and fiber atrophy. Examination of the slope change of the log-transformed EMG–force relation may help understand the selective loss of motor units or changes in muscle fibers that may affect motor unit distribution [40].

Finally, yet importantly, the limitations of the study should be acknowledged. In our simulation, many simplified assumptions were used. The primary focus of the model was the spatial distribution of small and large motor units. Although a significant difference in the slope of the log-transformed EMG–relation was confirmed in the simulation by solely varying motor unit spatial distribution, there are many factors that can influence the EMG–force relation (the *b* values). For example, variations in muscle fiber length, conduction velocity, arm geometry, surface electrode size, and configuration may all have an effect on recorded surface EMG signals [41–43]. These physiological and recording factors should be considered for data analysis and interpretation. Furthermore, surface EMG has an intrinsic limitation in that the electrode can only record a finite portion of the muscle, notably less in larger muscles. Therefore, the experimental EMG recorded here only reflects activity in a portion of the muscle and is not necessarily applicable to the whole muscle.

5. Conclusions

This study examined the spatial dependence of EMG–force relations using both simulated and experimental approaches. In the simulation study, it was found that the slope (*b*) of the log-transformed EMG–force relation was sensitive to three different motor unit depth distributions. In the experimental study, the slope distribution of the log-transformed EMG–force relation in the biceps brachii muscle showed a spatial dependence along the proximal–distal direction. The findings of this study indicate that the slope of the log-transformed EMG–force relation can provide a convenient and useful adjunct parameter in the investigation of muscle or motor unit changes associated with pathology, injury, and aging.

Author Contributions: Conceptualization, C.S.K. and P.Z.; methodology, C.H., M.C. and Z.L.; software, C.H., M.C. and Z.L.; validation, C.S.K. and P.Z.; formal analysis, C.H.; investigation, C.H., M.C., Z.L., C.S.K. and P.Z.; resources, P.Z.; data curation, C.H.; writing—original draft preparation, C.H.; writing—revision, review and editing, C.S.K. and P.Z.; visualization, C.H. and P.Z.; supervision, P.Z.; project administration, C.H. and P.Z.; All authors have read and agreed to the published version of the manuscript.

Funding: This work was supported in part by Shandong Provincial Natural Science Foundation (ZR2021QH053, ZR2021QH267, ZR2020KF012), and the National Natural Science Foundation of China (82102179).

Institutional Review Board Statement: The study was conducted in accordance with the Declaration of Helsinki and approved by the Ethics Committee of Guangdong Work Injury Rehabilitation Center (Approval Code: AF/SC-07/2015.28, Approval Date: 30 June 2015).

Informed Consent Statement: Informed consent was obtained from all subjects involved in the study.

Data Availability Statement: The data that support the findings of this study are available from the corresponding author upon reasonable request.

Conflicts of Interest: The authors declare no conflict of interest.

Abbreviations

au	arbitrary unit
EMG	electromyography
ISI	inter-spike intervals
MUAP	motor unit action potential
PFR	peak firing rate
RMS	root mean square amplitude
RTE	recruitment threshold

References

1. Manta, P.; Kalfakis, N.; Kararizou, E.; Vassilopoulos, D.; Papageorgiou, K. Distribution of muscle fibre types in human skeletal muscle fascicles: An autopsy study of three human muscles. *Funct. Neurol.* **1995**, *10*, 137–141. [PubMed]
2. Buchthal, F.; Schmalbruch, H. Motor unit of mammalian muscle. *Physiol. Rev.* **1980**, *60*, 90–142. [CrossRef] [PubMed]
3. Edgerton, V.R.; Smith, J.L.; Simpson, D.R. Muscle fibre type populations of human leg muscles. *Histochem. J.* **1975**, *7*, 259–266. [CrossRef] [PubMed]
4. Susheela, A.; Walton, J. Note on the distribution of histochemical fibre types in some normal human muscles: A study on autopsy material. *J. Neurol. Sci.* **1969**, *8*, 201–207. [CrossRef]
5. Elder, G.C.; Bradbury, K.; Roberts, R. Variability of fiber type distributions within human muscles. *J. Appl. Physiol.* **1982**, *53*, 1473–1480. [CrossRef] [PubMed]
6. Johnson, M.A.; Polgar, J.; Weightman, D.; Appleton, D. Data on the distribution of fbre types in thirty-six human muscles. An autopsy study. *J. Neurol. Sci.* **1973**, *18*, 111–129. [CrossRef]
7. Lexell, J.; Henriksson-Larsén, K.; Sjöström, M. Distribution of different fibre types in human skeletal muscles. 2. A study of cross-sections of whole m. vastus lateralis. *Acta Physiol. Scand.* **1983**, *117*, 115–122. [CrossRef] [PubMed]
8. Dahmane, R.; Djordjevič, S.; Šimunič, B.; Valenčič, V. Spatial fiber type distribution in normal human muscle: Histochemical and tensiomyographical evaluation. *J. Biomech.* **2005**, *38*, 2451–2459. [CrossRef]
9. Beck, T.W.; Housh, T.J.; Cramer, J.T.; Mielke, M.; Hendrix, R. The influence of electrode shift over the innervation zone and normalization on the electromyographic amplitude and mean power frequency versus isometric torque relationships for the vastus medialis muscle. *J. Neurosci. Methods* **2008**, *169*, 100–108. [CrossRef]
10. Herda, T.J.; Zuniga, J.M.; Ryan, E.D.; Camic, C.L.; Bergstrom, H.C.; Smith, D.B.; Weir, J.P.; Cramer, J.T.; Housh, T.J. Quantifying the effects of electrode distance from the innervation zone on the electromyographic amplitude versus torque relationships. *Physiol. Meas.* **2013**, *34*, 315–324. [CrossRef]
11. Bhadane, M.; Liu, J.; Rymer, W.Z.; Zhou, P.; Li, S. Re-evaluation of EMG-torque relation in chronic stroke using linear electrode array EMG recordings. *Sci. Rep.* **2016**, *6*, 28957. [CrossRef] [PubMed]
12. Fuglevand, A.J.; Winter, D.A.; Patla, A.E. Models of recruitment and rate coding organization in motor-unit pools. *J. Neurophysiol.* **1993**, *70*, 2470–2488. [CrossRef] [PubMed]
13. Zhou, P.; Rymer, W.Z. Factors Governing the Form of the Relation Between Muscle Force and the EMG: A Simulation Study. *J. Neurophysiol.* **2004**, *92*, 2878–2886. [CrossRef] [PubMed]
14. Zhou, P.; Suresh, N.L.; Rymer, W.Z. Model Based Sensitivity Analysis of EMG–Force Relation with Respect to Motor Unit Properties: Applications to Muscle Paresis in Stroke. *Ann. Biomed. Eng.* **2007**, *35*, 1521–1531. [CrossRef]
15. Huang, C.; Chen, M.; Zhang, Y.; Li, S.; Zhou, P. Model-Based Analysis of Muscle Strength and EMG-Force Relation with respect to Different Patterns of Motor Unit Loss. *Neural Plast.* **2021**, *2021*, 5513224. [CrossRef]
16. Robertson, J.W.; Johnston, J.A. Modifying motor unit territory placement in the Fuglevand model. *Med. Biol. Eng. Comput.* **2017**, *55*, 2015–2025. [CrossRef]
17. Herda, T.J.; Weir, J.P.; Ryan, E.D.; Walter, A.A.; Costa, P.B.; Hoge, K.M.; Beck, T.W.; Stout, J.R.; Cramer, J.T. Reliability of absolute versus log-transformed regression models for examining the torque-related patterns of response for mechanomyographic amplitude. *J. Neurosci. Methods* **2009**, *179*, 240–246. [CrossRef]
18. Herda, T.J.; Walter, A.A.; Costa, P.B.; Ryan, E.D.; Stout, J.R.; Cramer, J.T. Differences in the log-transformed electromyographic–force relationships of the plantar flexors between high- and moderate-activated subjects. *J. Electromyogr. Kinesiol.* **2011**, *21*, 841–846. [CrossRef]
19. Eason, T.; Gavel, C.R.; Hawley, K.A.; Galen, S.S.; Malek, M.H. Reliability of the log-transformed EMG amplitude-power output relationship for incremental knee-extensor ergometry. *Muscle Nerve* **2015**, *52*, 428–434. [CrossRef]
20. De Luca, C.J.; Hostage, E.C. Relationship Between Firing Rate and Recruitment Threshold of Motoneurons in Voluntary Isometric Contractions. *J. Neurophysiol.* **2010**, *104*, 1034–1046. [CrossRef]
21. De Luca, C.J.; Contessa, P. Biomechanical benefits of the onion-skin motor unit control scheme. *J. Biomech.* **2015**, *48*, 195–203. [CrossRef] [PubMed]
22. Piotrkiewicz, M.; Türker, K.S. Onion Skin or Common Drive? *Front. Cell. Neurosci.* **2017**, *11*, 2. [CrossRef]
23. Inglis, J.G.; Gabriel, D.A. Is the 'reverse onion skin' phenomenon more prevalent than we thought during intramuscular myoelectric recordings from low to maximal force outputs? *Neurosci. Lett.* **2021**, *743*, 135583. [CrossRef] [PubMed]
24. Erim, Z.; De Luca, C.J.; Mineo, K.; Aoki, T. Rank-ordered regulation of motor units. *Muscle Nerve* **1996**, *19*, 563–573. [CrossRef]
25. Oya, T.; Riek, S.; Cresswell, A.G. Recruitment and rate coding organisation for soleus motor units across entire range of voluntary isometric plantar flexions. *J. Physiol.* **2009**, *587*, 4737–4748. [CrossRef] [PubMed]
26. Jesunathadas, M.; Klass, M.; Duchateau, J.; Enoka, R.M. Discharge properties of motor units during steady isometric contractions performed with the dorsiflexor muscles. *J. Appl. Physiol.* **2012**, *112*, 1897–1905. [CrossRef]
27. Merletti, R.; Conte, L.L.; Avignone, E.; Guglielminotti, P. Modeling of surface myoelectric signals. I. Model implementation. *IEEE Trans. Biomed. Eng.* **1999**, *46*, 810–820. [CrossRef] [PubMed]
28. Lawrence, J.H.; De Luca, C.J. Myoelectric signal versus force relationship in different human muscles. *J. Appl. Physiol.* **1983**, *54*, 1653–1659. [CrossRef]

29. Knight, C.A.; Kamen, G. Superficial motor units are larger than deeper motor units in human vastus lateralis muscle. *Muscle Nerve* **2005**, *31*, 475–480. [CrossRef]
30. Stalberg, E. Macro EMG, a new recording technique. *J. Neurol. Neurosurg. Psychiatry* **1980**, *43*, 475–482. [CrossRef]
31. Farina, D.; Blanchietti, A.; Pozzo, M.; Merletti, R. M-wave properties during progressive motor unit activation by transcutaneous stimulation. *J. Appl. Physiol.* **2004**, *97*, 545–555. [CrossRef]
32. Mesin, L.; Merlo, E.; Merletti, R.; Orizio, C. Investigation of motor unit recruitment during stimulated contractions of tibialis anterior muscle. *J. Electromyogr. Kinesiol.* **2010**, *20*, 580–589. [CrossRef] [PubMed]
33. Liu, Y.; Chen, Y.-T.; Zhang, C.; Zhou, P.; Li, S.; Zhang, Y. Motor unit distribution and recruitment in spastic and non-spastic bilateral biceps brachii muscles of chronic stroke survivors. *J. Neural Eng.* **2022**, *19*, 046047. [CrossRef] [PubMed]
34. Tang, A.; Rymer, W.Z. Abnormal force—EMG relations in paretic limbs of hemiparetic human subjects. *J. Neurol. Neurosurg. Psychiatry* **1981**, *44*, 690–698. [CrossRef] [PubMed]
35. Gemperline, J.J.; Allen, S.; Walk, D.; Rymer, W.Z. Characteristics of motor unit discharge in subjects with hemiparesis. *Muscle Nerve* **1995**, *18*, 1101–1114. [CrossRef]
36. Zhou, P.; Li, X.; Rymer, W.Z. EMG-Force Relations During Isometric Contractions of the First Dorsal Interosseous Muscle After Stroke. *Top. Stroke Rehabil.* **2013**, *20*, 537–543. [CrossRef] [PubMed]
37. Suresh, N.L.; Concepcion, N.S.; Madoff, J.; Rymer, W.Z. Anomalous EMG–force relations during low-force isometric tasks in hemiparetic stroke survivors. *Exp. Brain Res.* **2015**, *233*, 15–25. [CrossRef]
38. Jahanmiri-Nezhad, F.; Hu, X.; Suresh, N.L.; Rymer, W.Z.; Zhou, P. EMG-force relation in the first dorsal interosseous muscle of patients with amyotrophic lateral sclerosis. *Neurorehabilitation* **2014**, *35*, 307–314. [CrossRef]
39. Li, L.; Hu, H.; Yao, B.; Huang, C.; Lu, Z.; Klein, C.S.; Zhou, P. Electromyography–Force Relation and Muscle Fiber Conduction Velocity Affected by Spinal Cord Injury. *Bioengineering* **2023**, *10*, 217. [CrossRef]
40. Lukács, M.; Vécsei, L.; Beniczky, S. Large motor units are selectively affected following a stroke. *Clin. Neurophysiol.* **2008**, *119*, 2555–2558. [CrossRef]
41. Lowery, M.; Vaughan, C.; Nolan, P.; O'Malley, M. Spectral compression of the electromyographic signal due to decreasing muscle fiber conduction velocity. *IEEE Trans. Rehabil. Eng.* **2000**, *8*, 353–361. [CrossRef] [PubMed]
42. Pasquet, B.; Carpentier, A.; Duchateau, J. Change in Muscle Fascicle Length Influences the Recruitment and Discharge Rate of Motor Units During Isometric Contractions. *J. Neurophysiol.* **2005**, *94*, 3126–3133. [CrossRef] [PubMed]
43. Del Santo, F.; Gelli, F.; Ginanneschi, F.; Popa, T.; Rossi, A. Relation between isometric muscle force and surface EMG in intrinsic hand muscles as function of the arm geometry. *Brain Res.* **2007**, *1163*, 79–85. [CrossRef] [PubMed]

bioengineering

MDPI

Article

Design and Analysis of a Deep Learning Ensemble Framework Model for the Detection of COVID-19 and Pneumonia Using Large-Scale CT Scan and X-ray Image Datasets

Xingsi Xue, Seelammal Chinnaperumal, Ghaida Muttashar Abdulsahib, Rajasekhar Reddy Manyam, Raja Marappan and Sekar Kidambi Raju et al.

[1] Fujian Provincial Key Laboratory of Big Data Mining and Applications, Fujian University of Technology, Fuzhou 350011, China
[2] Department of Computer Science and Engineering, Solamalai College of Engineering, Madurai 625020, Tamil Nadu, India
[3] Department of Computer Engineering, University of Technology, Baghdad 10066, Iraq
[4] Amrita School of Computing, Amrita Vishwa Vidyapeetham, Amaravati Campus, Mangalagiri 522503, Andhra Pradesh, India
[5] School of Computing, SASTRA Deemed University, Thanjavur 613401, Tamil Nadu, India
[6] Department of Solar, Al-Nahrain Renewable Energy Research Center, Al-Nahrain University, Baghdad 64040, Iraq
* Correspondence: sekar_kr@cse.sastra.ac.in; Tel.: +91-9843710643

Abstract: Recently, various methods have been developed to identify COVID-19 cases, such as PCR testing and non-contact procedures such as chest X-rays and computed tomography (CT) scans. Deep learning (DL) and artificial intelligence (AI) are critical tools for early and accurate detection of COVID-19. This research explores the different DL techniques for identifying COVID-19 and pneumonia on medical CT and radiography images using ResNet152, VGG16, ResNet50, and DenseNet121. The ResNet framework uses CT scan images with accuracy and precision. This research automates optimum model architecture and training parameters. Transfer learning approaches are also employed to solve content gaps and shorten training duration. An upgraded VGG16 deep transfer learning architecture is applied to perform multi-class classification for X-ray imaging tasks. Enhanced VGG16 has been proven to recognize three types of radiographic images with 99% accuracy, typical for COVID-19 and pneumonia. The validity and performance metrics of the proposed model were validated using publicly available X-ray and CT scan data sets. The suggested model outperforms competing approaches in diagnosing COVID-19 and pneumonia. The primary outcomes of this research result in an average F-score (95%, 97%). In the event of healthy viral infections, this research is more efficient than existing methodologies for coronavirus detection. The created model is appropriate for recognition and classification pre-training. The suggested model outperforms traditional strategies for multi-class categorization of various illnesses.

Keywords: CT scan images; chest X-ray; transfer learning; COVID-19 detection; deep learning

Citation: Xingsi Xue, Seelammal Chinnaperumal, Ghaida Muttashar Abdulsahib, Rajasekhar Reddy Manyam, Raja Marappan and Sekar Kidambi Raju et al. Design and Analysis of a Deep Learning Ensemble Framework Model for the Detection of COVID-19 and Pneumonia Using Large-Scale CT Scan and X-ray Image Datasets. *Bioengineering* **2023**, *10*, 363. https://doi.org/10.3390/bioengineering10030363

Academic Editor: Enrique Berjano

Received: 11 February 2023
Revised: 9 March 2023
Accepted: 10 March 2023
Published: 16 March 2023

1. Introduction

CT scans, medical imaging, or chest X-rays have been proposed as a viable method for detecting COVID-19 quickly and early. CT scan has demonstrated high sensitivity in identifying COVID-19 at the initial assessment of patients. In severe circumstances, it may efficiently correct RT-PCR false negatives [1–4]. Nevertheless, this is a complex and time-consuming process, because a professional must interpret the X-rays and CT imaging to establish if an individual is COVID-19 positive. The first diagnostic tool to be used to identify COVID-19 pathology is a chest X-ray. A single chest X-ray image is insufficient for reliable prediction and treatment of COVID-19 [5–8]. To address this limitation, multiple medical data findings are combined, and predictive classifications are generated, leading to improved accuracy compared to using only a single test image. However, extracting relevant features from COVID-19 images is challenging due to daily fluctuations in characteristics and differences between cases [9–12]. Many conventional machine learning (ML) techniques have already been applied to automatically classify digitized chest data. Using an SVM classification model, frequent patterns were generated from the pulmonary surface to distinguish between malignant and benign lung nodules. Backpropagation networks have been used to categorize imagery as normal or malignant using a gray-level co-occurrence matrix technique [13–15].

Convolutional neural networks (CNNs) can remove valuable features in picture categorization tasks. This feature extraction is performed with transfer learning. Pre-trained approaches collect the general properties of large-scale data such as ImageNet and then apply them to the task. DL algorithms are better than traditional NN models if enough labeled images are available. The CNN model is one of the most common DL algorithms for diagnostic imaging, with excellent results. Despite ML algorithms, the effectiveness of CNNs depends on acquiring feature representation from property imagery. As a result, the obtainability of pre-trained models such as DenseNet, ResNet, and VGG-16 is extremely helpful in this procedure and appears to be extremely interesting for COVID-19 identification using chest CT images and X-ray images. Transferring acquired knowledge from a pre-trained network that has completed one function to a new task is a systematic approach for training CNN architecture. This technique is faster and more efficient because it does not require a large annotation training dataset; as a result, most academics, particularly in the medical field, prefer it.

Transfer learning can be performed in three ways [16–20] as follows: shallow tuning adjusts individuals from the previous classification layer to novel problems while leaving the limitations of the remaining levels untrained; the deep tune function is used to retrain the variables that were before the network from end to end; fine tuning intends to prepare many layers after layer and gradually adjust to learning variables until a critical performance is obtained. Knowledge transfer in detecting X-ray images through an impressively executed fine-tuning procedure.

The motivational aspect of the proposed model, along with the significant contributions, are as follows. The suggested system uses transfer learning to train a model faster. The training models compute the appropriate weights. Tuning in to the work at hand, it is identifying COVID-19. The techniques are stacked to anticipate output class. The meta-model in this system is an individual neuron that correctly forecasts the input category using the outputs. To train a model faster, the proposed system employs a learning approach. The weights of the pre-trained method are correctly adjusted during training, and the approaches are integrated via stacking. The best model architecture and training parameters for COVID-19 and pneumonia diagnosis should be automated in order to produce better results. Using transfer learning methodologies solves research gaps and reduces training time.

The research is organized as follows: Section 2 focuses on the related existing works for COVID-19 detection cases; Section 3 discusses the proposed work and the details of the algorithms presented in this study; Section 4 illustrates the graphical plots and tabular representation for the experimentation of the proposed and existing methods; Section 5 summarizes the proposed work and concludes with its enhancements.

2. Literature Survey and Related Work

The unique stacked ensemble model is proposed for diagnosing coronavirus from a person's chest CT scans or chest X-ray imaging [21]. The suggested model is a layered ensemble before heterogeneous ML models. Some of the transfer learning techniques are VGG-19, ResNet, DenseNet, and fully connected networks are the pre-trained DL models. By adjusting the extra fully connected tiers in each of them before the model, prospective candidates for the classification algorithm are obtained. After an extensive search, three outstanding models were chosen to create approximate weighting heterogeneity stacked ensembles. Early diagnosis of coronavirus-positive cases prevents the disease from spreading further in the population and allows for earlier treatment of the general public [22]. CT imaging and chest X-ray images have recently shown distinctive characteristics that indicate the severity of COVID-19 in the lungs. The advancement of AI technology, specifically DL-based healthcare, is crucial in handling vast amounts of data to achieve accurate and quick outcomes in diagnostic imaging. It can also aid in the more efficient and precise diagnosis of pathogens while assisting in remote locations. Various two-dimensional convolution techniques have been proposed for analyzing chest X-ray images to identify COVID-19 with a precision of 99% and a validation accuracy of 98%, with a loss of around 0.15%.

Rapid identification of coronavirus [23] was a challenging and time-consuming task that computer-aided diagnosis (CAD) techniques have effectively addressed, enabling quick decision-making and meaningful tasks for surveillance, treatment, and medication. The use of chest X-ray (CXR) technique has become a cost-effective and practical alternative to CT scanning and real-time polymerase chain reaction (RT-PCR) testing, which is the most widely used detection method for COVID-19. Although several CAD approaches have been developed for detecting COVID-19, various issues have hindered their effectiveness. This research develops a deep CNN (DCNN) model using ResNet32 with the convolutional block attention module to detect COVID-19 and pneumonia.

VGG-16 surpasses all these models on the dataset with an accuracy of 85.33% [24]. Clinicians employing RT-PCR tests can diagnose COVID-19, but it has a high rate of false positive (FP) and false negative (FN) findings, and it takes a long time. One way to increase the true positive (TP) ratio is to run many experiments simultaneously. CT scans and X-ray images can be used to identify COVID-19-related pneumonia earlier. Upwards of 95% efficiency can be reached using current DL methods. Various CNN (CovNet)-based DL models are applied to diagnose pneumonia, including ResNet 152 v2, InceptionResNet v2, Xception, Inception v3, ResNet 50, NASNetLarge, DenseNet 201, and VGG 16 [25].

A DL framework can be created to assist in evaluating CT tests to offer diagnostics that can help save time in managing the disease. In this study, a DL algorithm was adapted to identify COVID-19 using extracted features from chest X-rays and CT images. Many transfer learning models were initially used and compared, and a VGG-19 model was subsequently tweaked to produce the best results for disease diagnosis. The diagnosis accuracy of all models was evaluated using a sample of 1000 images. The VGG-19 model has higher precision, sensitivity, and specificity, with 99.4% precision, 97.4% sensitivity, and 99.4% specificity in the timely identification of coronavirus. ML and image analysis performed well [26]. Researchers present an approach to examining the possibility of deep transfer learning in developing a classification to identify coronavirus-positive patients utilizing CT and CXR imaging in this research. The preprocessing data method expands the training datasets to lessen overfitting and improve the model's generalization capacity. To use a preprocessing data method, we evaluated pre-trained DL models: ResNet50, InceptionV3, VGGNet-19, and Xception. DL is better at finding coronavirus cases, according to the research. The VGGNet-19 model surpasses the other three models presented using the CT image dataset when each modality is considered separately, according to the findings of the tests [27]. It is true that Deep Neural Networks (DNNs) have made significant progress in computer vision tasks, including medical imaging, and DenseNet has been evaluated for categorizing COVID-19 chest X-ray images. The COVID-19 pandemic has highlighted the importance of medical imaging in the diagnosis and monitoring of diseases, and machine learning algorithms such as DNNs can help automate and streamline the analysis process. DenseNet is a popular convolutional neural network architecture that has shown promising results in various image classification tasks. A public database of 6432 chest X-ray images divided into three classes was also used. The three DenseNet modeling variants, DenseNet121, DenseNet169, and DenseNet201, were trained via transfer learning and fine tuning. After analyzing the results, it was found that DenseNet201 had the best accuracy rate of 0.9367 and the most negligible validation loss of 0.1653 for COVID-19 classifications in chest X-ray images [27,28].

Patients must wait an extended period of time to receive the results of blood tests to detect COVID-19. This method uses a DL technique to detect COVID-19 quickly. DL and neural algorithms logistic regression is a statistical tool for predicting the outcome of radiology data, such as CT scans and X-ray images, that are fed into these algorithms. Using this approach, positive instances of COVID-19 will be recognized more quickly [29]. This research focuses on building a DL approach for detecting and diagnosing COVID-19 utilizing chest CT scans. A public database was employed for this, which included 349 CT scans of 216 patients with COVID-19 clinical abnormalities and CT scans of 397 healthy people. Accuracy, precision, recall, Matthews coefficients correlation (MCC), and F-measure criteria were used to evaluate diagnostic accuracy. A 10-fold cross-validation procedure was used to assess the reliability of the approach. CNN has an accuracy rate of 92.63%, a precision of 92.95%, a recall of 93.18%, an MCC of 85.2%, and an F1 measure of 93.06%. Based on the results achieved with the approach taken within the focus of this research, the mentioned approach could be used as a supplement or replacement for traditional therapeutic tests in coronavirus outbreaks [29,30]. For the identification and classification of COVID-19, a transfer learning method with precise adjustment was used in this study. VGG16, DenseNet-121, ResNet-50, and MobileNet were employed as before models. DL models were developed using a dataset of 7232 (COVID-19 vs. healthy) chest X-ray images. A local dataset of 450 chest X-ray images from Pakistani patients was gathered or used for tests and predictions. Many essential characteristics, such as recall, specificity, F1-score, accuracy, loss graphs, and confusing matrix, were generated to verify the accuracy of the algorithms. VGG16, ResNet-50, DenseNet-121, and MobileNet attained an accuracy of 83.27%, 92.48%, 96.49%, and 96.48%, respectively [31].

This research aimed to show how DL, ML, and image processing can help with the fast and accurate identification of COVID-19 from two of the most frequently used neuroimaging techniques, chest X-rays and CT scans. CNN models based on the Alexnet model were proposed for CAD of coronavirus from CT and X-ray images. The effectiveness of this technique was tested on the obtained dataset. COVID-19 CT and X-ray scans were categorized, and CT scans were used to track the progression of the disease [32]. COVID-19 has been successfully identified by utilizing CT scans and X-rays to assess lung images. However, it takes a team of radiologists and a lot of time to review each report, which is time-consuming. As a result, this work proposes an automatic COVID-19 recognition and categorization algorithm based on DNNs. The provided method performs pre-processing, extraction of features, and classifications. The input image is processed using median filtering at an earlier stage. The VGGNet-19 model, built on a CNN, is then used as a feature representation. Finally, to identify and classify the presence of COVID-19, an ANN is used as a classification algorithm [33]. Researchers worldwide are investigating new ML approaches, such as DL, to help medical experts diagnose COVID-19 disease using medical information such as X-ray films and CT scans. ML techniques for demonstrating the first COVID-19 identification from chest X-rays would have been effective in controlling this epidemic, because the facility for chest X-rays is available even in smaller cities and is significantly less expensive. As a result, researchers suggest a CNN for identifying the presence and absence of COVID-19. The CNN model is compared with the other models based on ancient and transfer learning. Compared to ML algorithms, the suggested CNN is more efficient (KNN, SVM, DT, etc.) [34,35]. COVID-19 may be identified using DL techniques, a crucial task for treatment possibilities identified based on data these days. The development of AI, ML, DL, and diagnostic imaging approaches offers excellent performance, particularly in the areas of detection, classification, and segmentation. The advances in imaging technologies that have allowed clinicians to monitor the body and accurately diagnose and assess patients without surgery. It also highlights the use of several imaging technologies, including CT with a DL model built on a CNN, for detecting COVID-19 [36–43].

3. Methods

This section focuses on the need for the proposed system model in detail.

3.1. Need for Proposed Model

The proposed model is required to fulfill the following requirements: to obtain better measures, and to automate optimum model architecture and training parameters in COVID-19 and pneumonia diagnosis; to apply the transfer learning approaches to solve content gaps and shorten training duration; to perform multi-class classification for X-ray imaging tasks by applying an upgraded VGG16 deep transfer learning; to minimize the complexity of the layered design to perform the multi-class categorization of various illnesses.

3.2. System Model

This section explores the proposed model as sketched in Figure 1. The spatial domain procedure is applied to the image pixels and altered pixel by pixel in this filtering. It is a filtering mask that shifts from one pixel to another by performing several operations. It will remove noise from the image by smoothing it.

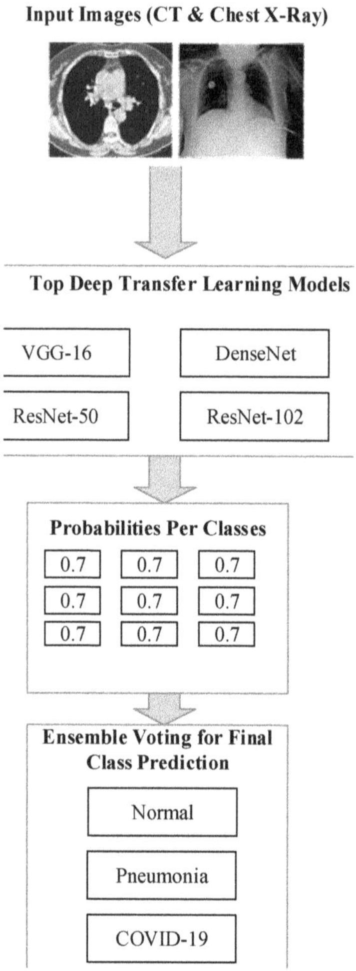

Figure 1. The architecture of the proposed model.

The spatial domain filtering is divided into linear and non-linear filters. The mean and Wiener filters are the most often used linear filtering techniques, whereas the median filter is a non-linear filtering approach. The mean value in the mean filter is derived from the computation of neighbors and center pixel values in the N × N size. It is calculated after calculating the center pixel value as follows:

$$Y(M,N) = Mean\ \{X[i,j],\ [i,j] \epsilon\ w\} \tag{1}$$

where w are the pixel positions in the neighborhood.

The Wiener filter is applied for filtering the consistent pixel values that define the constant power additive noise. This filtering technique is used for adaptive filtering of the image pixel-wise. Two variations are computed by computing the pixels in the neighborhood with the following mean and standard deviation:

$$\mu = \frac{1}{N \times M} \sum\nolimits_{(n_1, n_2 \in \eta)} a(n_1, n_2) \tag{2}$$

$$\sigma^2 = \frac{1}{N \times M} \sum\nolimits_{(n_1, n_2 \in \eta)} a^2(n_1, n_2) - \mu^2 \tag{3}$$

where η represents the $N \times M$ of the current pixel; with this estimation, the pixel-wise Wiener filter was applied over the denoised image, which is computed as follows:

$$b(n_1,\ n_2) = \mu + \frac{\alpha^2 - v^2}{\alpha^2}(a(n_1, n_2) - \mu) \tag{4}$$

where v^2 represents the noise variance.

The median filter works by the pixels in the window sorted in ascending order. The median value of the $N \times M$ image is changed based on the central pixel values using

$$Y(M,N) = Median\ \{X[i,j],\ [i,j] \epsilon\ w\} \tag{5}$$

where w is the pixels of the neighborhood.

Image enhancement uses quality enhancement techniques to improve the image contents based on essential features, including intensity, edges, and corners. Histogram equalization redistributes the pixel values of an image to increase contrast, making it easier to distinguish between different image features. This technique can help to reduce the number of pixels that are saturated or redundant in certain areas of the image, but its primary goal is to enhance the overall visual quality of the image. This process increases the contrast level of the image and the computation speed of content-based medical image retrieval. Histogram equalization is a technique that improves the image's visual quality by adjusting the noisy and blurred pixel values. The histogram represents the discrete function or the intensity distribution with the graphical format applied to dark images. The process of histogram equalization for contrast enhancement is defined as follows:

1. Read the input image.
2. Compute the histogram, probability of each pixel, and probability density function of the input image.
3. Equalize the histogram.

The histogram represents the image's intensities using bars representing pixel frequency. For a given input image, the histogram is computed as follows:

$$H(r_k) = n_k \tag{6}$$

where r_k is the kth gray level value ranging from 0 to L-1, and n_k is the number of pixels in the image which have the gray level in r_k. Then, determine the probability of the gray level using

$$p(r_k) = \frac{n_k}{M \times N} \tag{7}$$

where n_k is the number of pixels, $M \times N$ defines the image size with M rows and N columns. The pdf of the image is computed using

$$p_k = \sum_{i=0}^{k} n_k \times \frac{1}{M \times N} \tag{8}$$

Then, the histogram equalization is applied to enhance the quality of the image by correcting its intensity values for the discrete case using

$$S_k = (L-1) \times p_k(r) \tag{9}$$

where $p_k(r)$ represents the pdf.

3.3. Fast Attention-Based ResNet

Attention-based ResNet is used to perform feature extraction and classification. The architecture of the proposed model is depicted in Figure 2. The learning rate of the proposed model is set to 0.0001, and the number of epochs performed for classification is 10. The extracted set of features has many irrelevant and redundant features that need to be removed because it reduces the accuracy of the process. Then, the feature selection is performed to remove the irrelevant and redundant characteristics [7]. The attention module is responsible for the extraction of features by learning the weights of the features corresponding to the scaling attacks. The output of the attention layer is computed using

$$Atn_i(Z) = Q_i(Z).\ F_i(Z) + F_i(Z) \tag{10}$$

where $Q_i(Z)$ denotes the attention weight and $F_i(Z)$ denote the features. In the attention layer, the relationship between the features is computed to achieve more relevant information from the features, which can be computed as,

$$R(F_1;F_2) = \int_{F_1}^{n}\int_{F_2}^{n} p(F_1,F_2)\log\frac{p(F_1,F_2)}{p(F_1)p(F_2)}dF_1dF_2 \tag{11}$$

where $p(F_1,F_2)$ denote the function of probability between the features F_1 and F_2 and $p(F_1)$, $p(F_2)$ denote the individual function of marginal density, respectively. The inception layer learns specific features deeply by reducing the size of the initial patch, which affects the original information required for the classification process. Based on the selected features, classification is carried out. This process ascertains whether a given unknown packet is standard or malicious. The softmax layer performs this classification, in which the cross entropy is computed to determine the output loss, which can be calculated as

$$Loss\ (x) = -\sum_k \log(st(x)_k)q_i \tag{12}$$

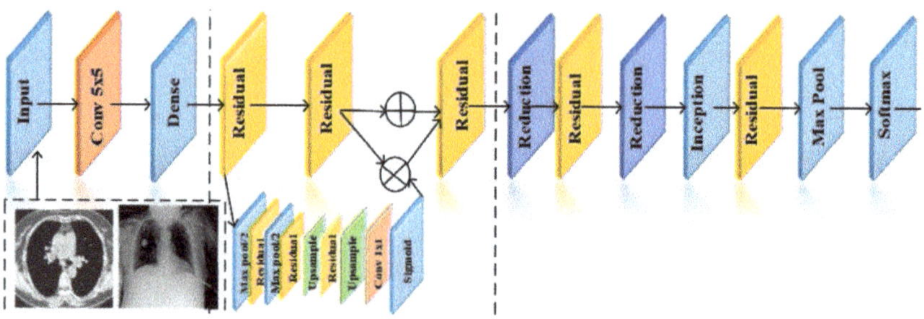

Figure 2. Attention-based ResNet.

Formulate the softmax $st(x)$ of a vector using

$$st(x)_k = \frac{e^{x_k}}{\sum_m e^{x_m}} \tag{13}$$

The classification is performed by increasing the score corresponding to the input images.

3.4. Enhanced VGG-16 Architecture for X-ray Images

Figure 3,the VGG-16 architecture is implemented to extract and classify the features using DL [8]. Initially, the features of the segmented region are extracted. The proposed VGG-16 has three layers: a convolutions layer, a fully connected layer, and a softmax layer. VGG 16 extracts low-level features with small masks and has fewer layers than VGG 19.

The max pooling and average pooling layers extract the features from the image. These layers extract the features from the segmented region, as shown in Table 1. The convolution layer concatenates the pooling results using the sigmoid function(δ) with the measure of

$$C(F) = \delta\left(f^{7\times7}[F_A; F_M]\right) \tag{14}$$

where $F_A \in r^{1\times h\times w}$ and $F_M \in r^{1\times h\times w}$ represent expected and max pooling measures.

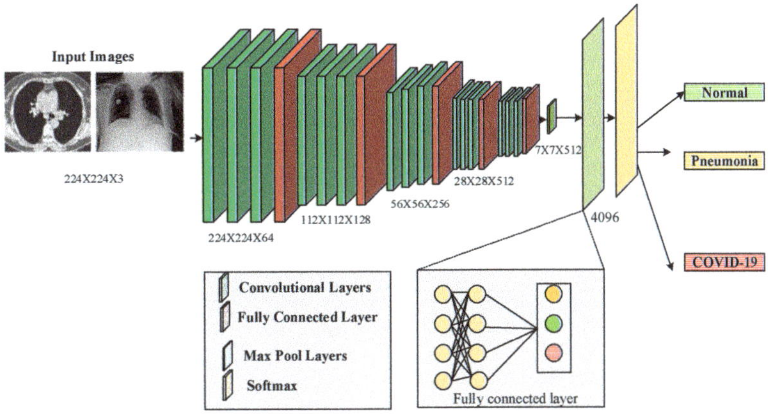

Figure 3. The architecture of VGG 16.

Table 1. Features and Aesthetics of the model.

Color	Standard deviation, intensity, kurtosis, skewness
Texture	Homogeneity, entropy, correlation
Shape	Artery—radius, length, angle, area, eccentricity

The fully connected layer receives the extracted results using dense, flattened, and dropout layers to bring the final classification as normal or abnormal using

$$C(p = q/s) = \frac{e^b}{\sum_j e^{b_j}} \tag{15}$$

Algorithm 1 presents the segmentation process of dividing an image into multiple segments or regions based on some criteria, such as color, texture, or shape. VGG16 is a convolutional neural network that has been widely used for image classification tasks. However, it can also be used for feature extraction and segmentation tasks.

Algorithm 1: VGG 16

Input: Segmented Region (S.R.)
Output: Normal (N) or abnormal (AN)
{
Begin
Initialize $\{fc, fs, ft\}$
$fc \leftarrow \{fc1, fc2, ..fcn\}$
$fs \leftarrow \{fs1, fs2, ..fsn\}$
$ft \leftarrow \{ft1, ft2, ..ftn\}$
Initialize the feature extraction data.
for $i \leftarrow o \; to \; n$ **do**
Extract fc from S.R.
Extract fs from S.R.
Extract ft from S.R.

Algorithm 1: *Cont.*

$F \leftarrow \{fc, fs, ft\}$
Extract the features by average pooling layer F_A
Extract the features by maxpool layer F_M
Concatenate the features
Classifying the images using softmax layer
Class $\leftarrow \{\ N,\ AN\}$
end for
}
return class
end
}

4. Results and Discussion

This section analyzes the experimental results of the proposed method.

4.1. Dataset

Five different databases are applied to evaluate the performance of the proposed method. Two databases provide chest X-ray imaging, while the remainder include chest CT scans [11,12]. Every database was divided into three parts: test, validation, and training set. The test set must have between at least 200 and 400 images to evaluate the model's flexibility well. The size of the test set determines the size of the verification set; the more significant the test set, the more prominent the testing set will be, and conversely. The other images were used to develop a training set. To perform the presented approach, the test and the validation test sets are organized, consisting of the same partitions of the positive and negative image samples. The hyper-parameters are tuned based on the training and test sets of the coronavirus images.

The descriptions of the datasets are given as follows:

The CT image dataset contains approximately 349 CT images from 216 patients with coronavirus and 397 patients without coronavirus. The images were collected from hospitals that treat both coronavirus and non-coronavirus patients, but only the positive and negative cases for coronavirus were included in the dataset [13,14]. Some of the samples of coronavirus images in both positive and negative classes are given in Figure 4. The features of the dataset are as follows:

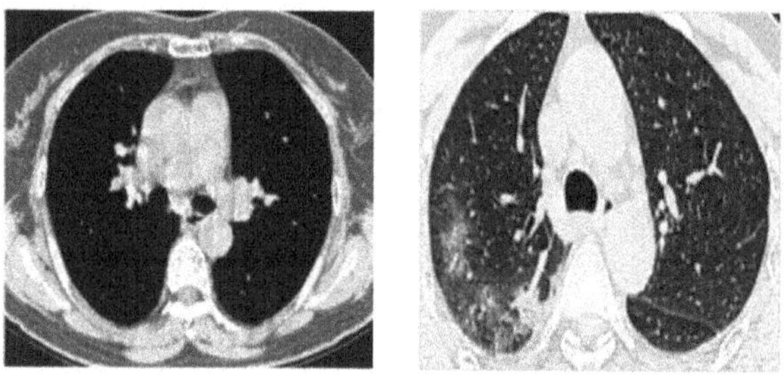

(a) (b)

Figure 4. (**a**) COVID-19 Negative (COVID-19 images collection dataset) (**b**) COVID-19 Positive (COVID-19 images collection dataset).

Types of Images: CT Images
Size of Dataset: 746 CT Scans
Positive Case images in Total: 349
Negative Case images in Total: 397
Validation Size Set: 118 Scans

Training Size Set: 425 Scans
Test Size Set: 203 Scans

The *COVID-19 images collection dataset* consists of images gathered from the public community, such as physicians and hospitals. The features of this dataset are as follows:

Types of Imaging: Chest X-rays
Size of Dataset: 579 Images
Training Size Set: 309
Validation Size Set: 70
Testing Size Set: 200
COVID-19 Positive Cases: 342
COVID-19 Negative Cases: 237

The *CT set COVID-19 dataset* consists of original CT scan imaging for 377 patients. A total of 15,589 CT scans are used for CT scan images of 95 coronavirus patients and 282 regular patients. The features of this dataset are as follows:

Types of Images: CT Scan Images
Size of Dataset: 12,058 Scans
Positive Images: 2282
Negative Images: 9776
Training Set: 11, 400
Validation Set: 258
Testing Set: 400

The COVID-19 radiography collection includes 1200 images of positive COVID-19, 1341 images of healthy patients, and 1345 images of patients with fever infection. This dataset has the following characteristics:

Figure 5 CT scans, or computed tomography scans, use X-rays to create detailed images of the inside of the body. CT scans can be helpful in diagnosing respiratory illnesses such as SARS-CoV-2, the virus that causes COVID-19. CT scans can show the presence of lung abnormalities such as ground-glass opacities and consolidation, which can be indicative of viral pneumonia.

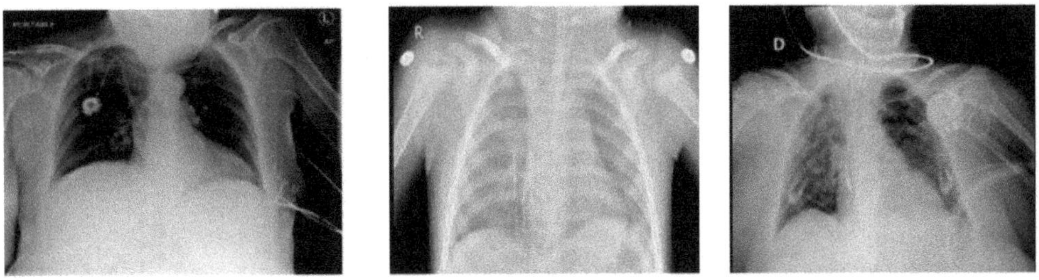

(a) (b) (c)

Figure 5. (**a**) Patient healthy image, (**b**) Patient with pneumonia, and (**c**) Patient with +Ve COVID-19 images COVID-19 negative.

Images Type: X-ray
Sum of Images: 2541
Negative Images: 2686
Positive Images: 1200
Training Size Set: 3086
Testing Size Set: 400
Validation Size Set: 400

SARS-CoV-2 CT scan images consists of 1252 CT scan images that are positive, whereas the remaining 1230 images are classified as non-infected viruses, with the following features:

Images Type: CT Scan Images
Size of Dataset: 2482 Images
Negative Images: 1230
Positive Images: 1252
Training Size Set: 1800
Testing Size Set: 400
Validation Set Size: 400

To mitigate the model overfitting issues of the training case, the training size is increased from 1275 images collected utilizing data augmentation based on random rotation, horizontal flipping, and color jittering.

4.2. Data Preprocessing and Augmentation Techniques

Using data augmentation methods, some enormous volumes of datasets must be tested using transfer learning or DL techniques [14,15]. Hence, this research model considers data augmentation methods with the datasets.

Resize or Crop by Random: This step represents the cropping of the input image to the unsystematic size of the image and the aspect ratio.
Rotation by Random: To perform this step, the sample is rotated by selecting an angle at random.
Horizontal Flip by Random: This step represents the flip action for the given input image randomly in a horizontal manner.
Color Jittering: This step randomly represents the modifications of the input image's contrast, saturation, and brightness.
Training Settings: The overall work is implemented through a Python framework, in which the variables are fixed for all experiments. The simulation parameters are shown in Table 2.

Table 2. Simulation parameters.

Parameters	Values
DL Model	Python (PyTorch library)
Number of Epochs	1 to 100
Learning Rate	0.003
Optimization Technique	ADAM
Loss	Cross Entropy
Size of Batch	16
Resized Cropping Scale in Random	0.5 to 1.0
Rotation Angle Range	(−5, +5)
Flipping Possibility Rate	0.5
Resized Size of Cropping in Random	224

4.3. Comparison Analysis

Classification accuracy is defined as the accuracy in performing the classification of both COVID-19 and viral pneumonia and computed using

$$Accuracy = \frac{T_rP + T_rN}{T_rP + T_rN + F_lP + F_lN} \times 100\% \quad (16)$$

It is important to find the optimal number of iterations that maximizes the model's accuracy without overfitting. This can be done by monitoring the validation accuracy during training, which measures how well the model performs on a separate set of data that is not used for training. The validation accuracy typically starts to plateau after a certain number of iterations, indicating that the model is no longer improving significantly and further training may lead to overfitting in Figure 6.

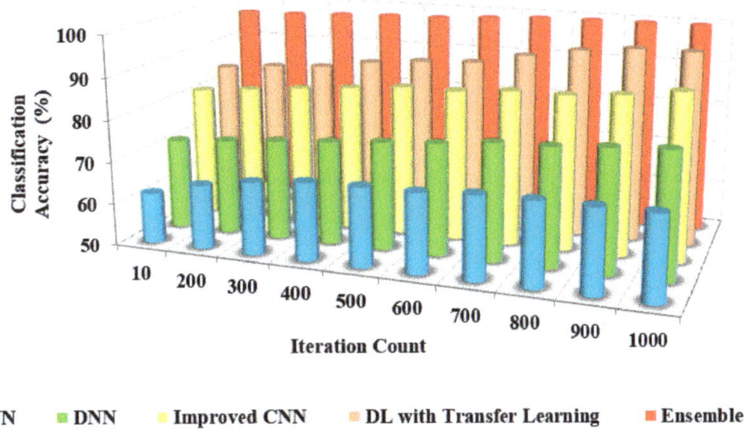

Figure 6. Iteration count vs. classification accuracy.

The greater the classification accuracy, the greater the efficiency of the approach. Figures 6 and 7 depict the relationship between the model's classification accuracy with others based on the iterations and the sample counts. The classification accuracy of the proposed ensemble model is high due to the extensive extraction of features and classification.

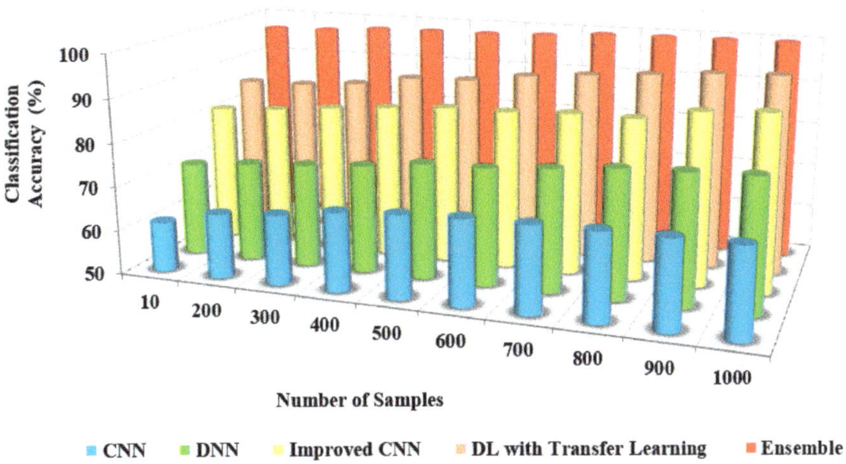

Figure 7. Number of samples vs. classification accuracy.

The feature set is generated from many features using a pre-trained model to eliminate redundant features [16,17]. The selection of transfer learning in the model is based on the recent study and literature analysis, contributing to increased classification accuracy. The existing approaches considered both necessary and redundant features for classification, which degraded the classification accuracy.

The classification accuracy of the proposed method with others based on users and iterations count is presented in Table 3. The classification and accuracy of the proposed method are 95% and 96%, respectively, whereas the existing approaches possess only about 62% to 72% of accuracy. The proposed method accurately classifies COVID-19 and non-COVID-19 images.

Table 3. Classification accuracy.

Methods	Classification Accuracy (%)	
	Iteration Count	No of Samples
CNN	61 ± 0.05%	62 ± 0.04%
DNN	71 ± 0.02%	72.6 ± 0.03%
Improved CNN (LeNet-5)	84 ± 0.03%	85.5 ± 0.04%
DL with Transfer Learning	92 ± 0.02%	92.5 ± 0.1%
Ensemble	95 ± 0.2%	96.2 ± 0.5%

The precision is the measure of relevancy in the classified images computed using

$$\text{Precision} = \frac{\text{T}_\text{r}\text{P}}{\text{T}_\text{r}\text{P} + \text{F}_\text{l}\text{P}} \times 100\% \tag{17}$$

The comparison of the precision of the presented method and the existing approaches to iteration count and the number of samples is illustrated in Figures 8 and 9. The figures show that the precision of the proposed model is high due to the implementation of spatial domain filtering, thereby eliminating the noise. Increasing the iteration count of a machine learning model can lead to improvements in recall, but this is not always the case Figure 10.

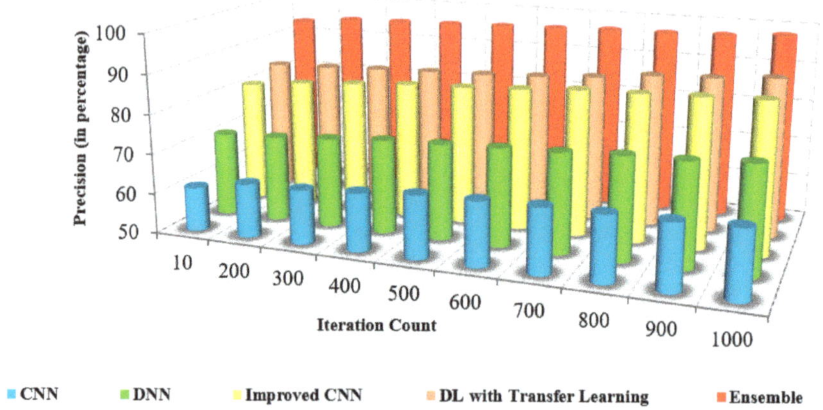

Figure 8. Iteration count vs. precision.

Further, the classification of the face and non-face regions was performed by utilizing the feature extraction in which the optimum approach performed the selection of features. The lack of consideration of pixel-based features and the removal of the noise level in the images of the existing approaches resulted in reduced precision [17,18]. The superiority of the proposed pre-trained approach is illustrated in Table 4, which presents the numerical comparison of the presented method and existing approaches in terms of the iteration count. The presented method's precision is 95% to 96%, whereas the existing approaches possess precision of about 61% to 71%. The proposed method produces better precision than other methods in detecting coronavirus.

The recall is called sensitivity, the measure of correctness between the classified images, calculated using

$$Recall = \frac{TP}{TP + FN} \tag{18}$$

Figures 11 and 12 sketches the comparison of recall measure with other methods over the fixed iterations in all cases of test data. The recall of the presented method is high due to the implementation of effective pre-processing techniques to eliminate the noise and correct the illumination.

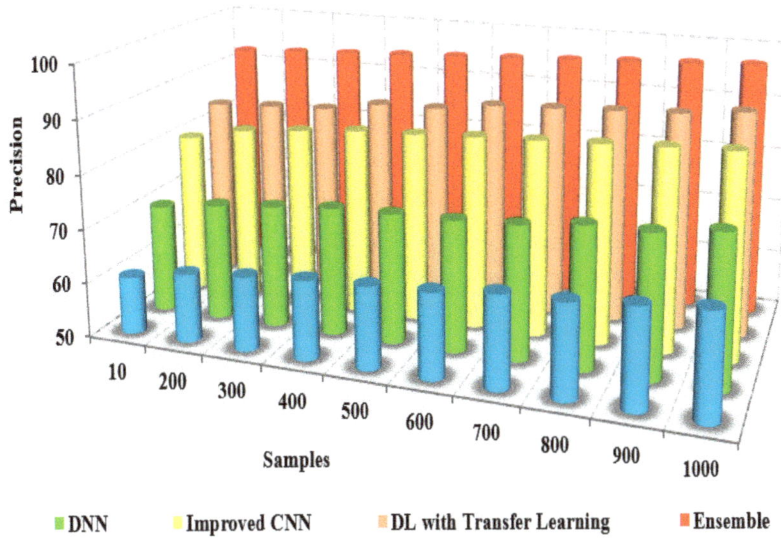

Figure 9. Samples vs. precision.

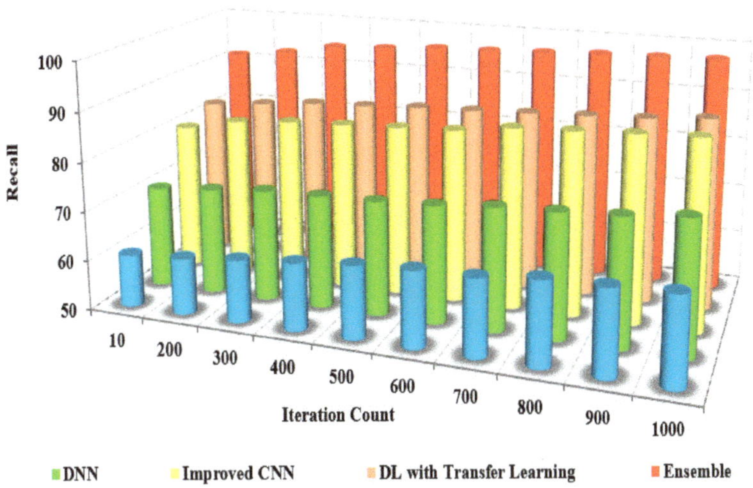

Figure 10. Iteration count vs. recall.

Table 4. Precision.

Methods	Precision (%)	
	Iteration Count	Number of Samples
CNN	60 ± 0.05%	62 ± 0.04%
DNN	71 ± 0.02%	72.6 ± 0.03%
Improved CNN (LeNet-5)	84 ± 0.03%	85.5 ± 0.04%
DL with Transfer Learning	92 ± 0.02%	92.5 ± 0.1%
Ensemble	95 ± 0.2%	96.2 ± 0.5%

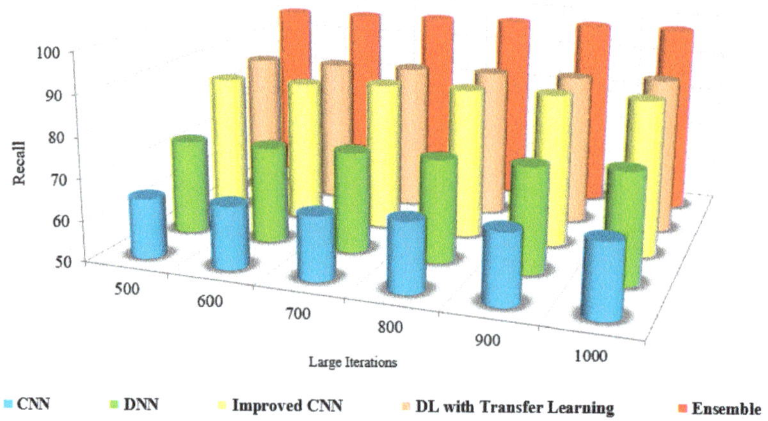

Figure 11. Large Iteration count vs. recall.

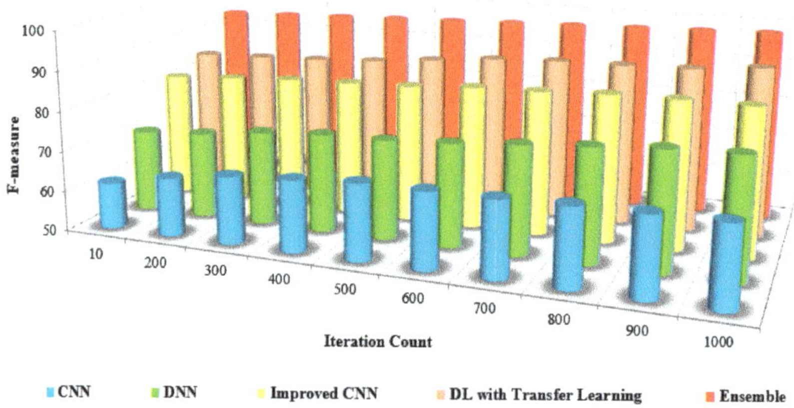

Figure 12. Iteration count vs. F-score.

Further, consideration of various features, such as low and high-level features, improved the recall results of the proposed model. The lack of noise elimination and consideration of integrated features restricted the recall of the existing approaches. The numerical analysis of recall with other methods is shown in Table 5. The presented method has a recall of 94% to 96%, whereas the existing approaches possess only about 62% to 72%. This leads to the conclusion that the presented method is more efficient than the existing approaches in performing coronavirus detection in healthy and viral infection cases [20].

Table 5. Recall.

Methods	Recall (%)	
	Iteration Count	**No. of Samples**
CNN	59.6 ± 0.05%	61.5 ± 0.04%
DNN	69.8 ± 0.02%	71.5 ± 0.03%
Improved CNN (LeNet-5)	84 ± 0.03%	83.5 ± 0.04%
DL with Transfer Learning	91.5 ± 0.02%	92 ± 0.1%
Ensemble	94.5 ± 0.2%	95.9 ± 0.5%

The F-measure is the combination of precision and recall calculated by computing the harmonic mean and measured using

$$F - measure = \frac{2 \times TP}{(2 \times TP + FP + FN)} \tag{19}$$

The comparison of the F-measure with other methods based on the iteration count is illustrated in Figure 13. The increased F-measure of the proposed method is due to the increase in precision and recall value.

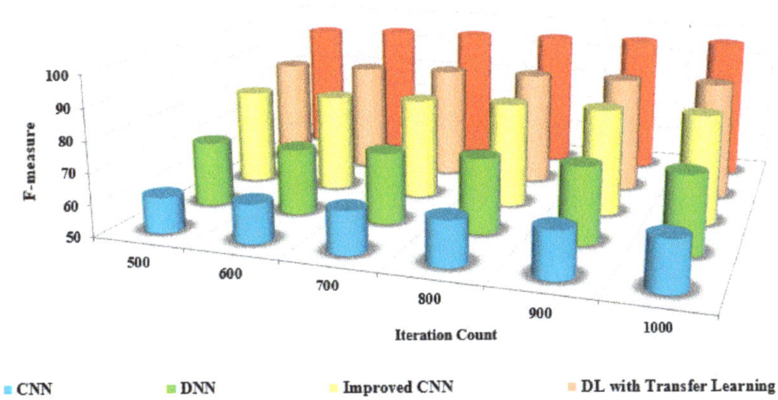

Figure 13. Large iteration count vs. F-score.

The detection and classification are carried out by pre-processing the input frames and extracting extensive features from the pre-processed images. The grouping of pixels is performed to determine the differentiation between the non-similar features. The classification of coronavirus and pneumonia is determined with an increased F-measure over the reduced F-measure of the other methods.

The efficiency of the presented method is proved by the numerical analysis presented in Table 6, which compares the F-measure with the other methods. The F-measure of the presented method is 95% to 97%, whereas the existing methods range from 63% to 72%. The model that was developed has been shown to be efficient for pre-training purposes in the detection and classification of masked faces.

Table 6. F-measure.

Methods	F-Measure (%)	
	Iteration Count	No. of Samples
CNN	60.2 ± 0.05%	62.3 ± 0.04%
DNN	71.2 ± 0.02%	72 ± 0.03%
Improved CNN (LeNet-5)	84.5 ± 0.03%	84.2 ± 0.04%
DL with Transfer Learning	92.2 ± 0.02%	92.5 ± 0.1%
Ensemble	95.2 ± 0.2%	96.7 ± 0.5%

The Receiver Operating Curve (ROC) depicts the relationship between the specificity and sensitivity in all possible cut-off values. The ROC curve is a graphical representation to visualize the efficiency of an approach. The better performance of an approach is depicted as the curve nearer to the left top corner of the ROC curve. The ROC curve of the proposed model with others is also analyzed for all the image sets of evaluation. The proposed model yields a better ROC curve than the other models in classification and detection [36,37]. The existing approaches possess reduced ROC curve values due to the inefficiency caused by image noise and the lack of feature integration for classification. The computational time is the time to compute specific tasks to obtain the desired result. Figure 13 compares the computational time of the proposed and other models based on the iteration factor. The computational time involved with the existing approaches is high due to the increased time consumption in training the model.

As the number of samples increases, the computation time required to process those samples also increases. This is because more samples typically mean more data to process and analyze, which can be computationally intensive. For example, if you're training a machine learning model on a dataset of images, the more images you have, the longer it will take to train the model Figure 14.

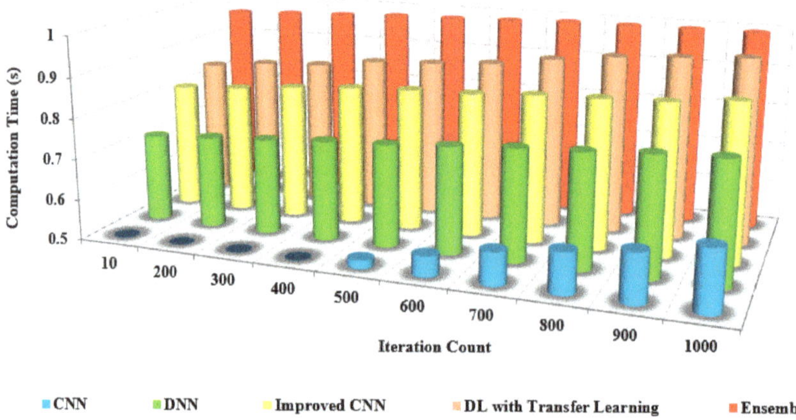

Figure 14. Iteration count vs. computation time.

The benefits of the proposed model are as follows: the pre-trained model possesses fewer layers than the existing models and thereby possesses low complexity. This characteristic of the proposed model requires a short computational time to detect and recognize the coronavirus.

Figure 15, this method provides minimal computation time without compromising the accuracy in detecting and recognizing the coronavirus [38,39]. The time complexity of the layered architecture runs in a linear, that is, O(n) time. The proposed model is computationally efficient and highly accurate for the multi-class categorization of distinct illness types [40–43].

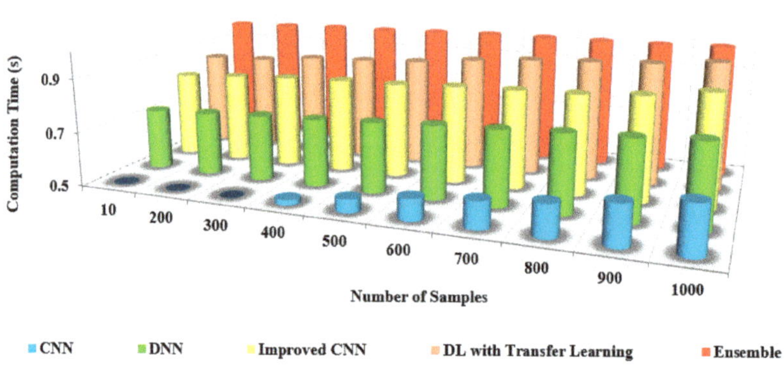

Figure 15. Number of samples vs. computation time.

Table 7 shows the comparison of the computation time with other models. From the table, it is clear that the computation time of our approach is shallow, at about 0.4 s. In contrast, the existing approaches possess an increased computation time of up to 0.9 s for both preprocessed and non-preprocessed.

Table 7. Computation time (s).

Methods	Computation Time	
	Masked Face	Unmasked Face
CNN	0.9	0.96
DNN	0.84	0.92
Improved CNN (LeNet-5)	0.78	0.84
DL with Transfer Learning	0.65	0.72
Ensemble	0.4	0.62

4.4. Limitations and Analysis

The limitations of the model are as follows: quality of outcomes based on the noise levels in the images affect the performance measures; integrating the relevant features are also an important requirement for the classification in the proposed model. The future research directions of the proposed model are as follows: enhancement using recent soft computing components and better recommender systems will be developed based on the features of datasets [19,20,44–48]; large datasets will be considered for validation to further improve performance measures using soft computing within the minimal computing time [25–27,49–63].

5. Conclusions

COVID-19 and pneumonia are usually encountered due to lung symptoms that can be identified through genes and studies. Early detection of coronavirus and effective management of its progression can be facilitated through the use of imaging tests. In identifying coronavirus disease, chest X-rays and CT are helpful imaging modalities. The wide availability of huge, annotated images based on transfer learning methods—VGG-16, ResNet, and DenseNet—has led to considerable progress in earlier transfer learning models for medical image classification. This ensemble learning technique is an accurate representation because it recovers a hierarchy of localized visual elements given the input. However, irregularities in the annotation of COVID-19 and pneumonia instances produced on X-ray imaging remain a much more problematic component of working on them. This research classified COVID-19 and pneumonia images in a vast chest X-ray and CT database using an ensemble learning system based on developments in the extraction features technique. The suggested methodology provided fast and thorough COVID-19 and pneumonia classification results and the ability to deal with data inconsistencies and a limited amount of class images. The classification and accuracy of the proposed method are about 95% and 96%, respectively, and the precision is found to be 95% to 96%. The recall is 94% to 96%, with an F-measure from 95% to 97%. The proposed method is more efficient than the existing approaches in performing coronavirus and pneumonia detection in healthy and viral infection cases. The devised model is adequate for pre-training in detection and classification. The time complexity of the layered architecture runs in a linear, that is, O(n) time, and takes less than 0.5 s. The proposed model is computationally efficient and highly accurate for the multi-class categorization of distinct illness types.

Author Contributions: Conceptualization, X.X.; Methodology, R.M.; Software, S.C.; Validation, G.M.A.; Formal analysis, R.R.M.; Writing—original draft, S.K.R.; Writing—review & editing, O.I.K. All authors have read and agreed to the published version of the manuscript.

Funding: This work is supported by the National Natural Science Foundation of China (No. 62172095), the Natural Science Foundation of Fujian Province (Nos. 2020J01875 and 2022J01644).

Institutional Review Board Statement: The study did not require ethical approval.

Informed Consent Statement: Informed consent was obtained from all subjects involved in the study.

Data Availability Statement: Not applicable.

Acknowledgments: I am writing to express my sincere gratitude for your support and generosity toward our research project. It is with great pleasure that I acknowledge the receipt of your funding support through the National Natural Science Foundation of China (No. 62172095) and the Natural Science Foundation of Fujian Province (Nos. 2020J01875 and 2022J01644).

Conflicts of Interest: The authors declare no conflict of interest.

References

1. Shah, F.M.; Joy, S.K.; Ahmed, F.; Hossain, T.; Humaira, M.; Ami, A.S.; Paul, S.; Jim, M.A.; Ahmed, S. A Comprehensive Survey of COVID-19 Detection Using Medical Images. *SN Comput. Sci.* **2021**, *2*, 434. [CrossRef] [PubMed]
2. Riaz, M.; Bashir, M.M.; Younas, I. Metaheuristics based COVID-19 detection using medical images: A review. *Comput. Biol. Med.* **2022**, *144*, 105344. [CrossRef] [PubMed]
3. Liang, S.; Liu, H.; Gu, Y.; Guo, X.; Li, H.; Li, L.; Wu, Z.; Liu, M.; Tao, L. Fast automated detection of COVID-19 from medical images using convolutional neural networks. *Commun. Biol.* **2021**, *4*, 35. [CrossRef] [PubMed]
4. Singh, M.; Bansal, S.; Ahuja, S.; Dubey, R.K.; Panigrahi, B.K.; Dey, N. Transfer learning–based ensemble support vector machine model for automated COVID-19 detection using lung computerized tomography scan data. *Med. Biol. Eng. Comput.* **2021**, *59*, 825–839. [CrossRef]
5. Das, A.; Ghosh, S.; Thunder, S.; Dutta, R.; Agarwal, S.; Chakrabarti, A. Automatic COVID-19 detection from X-ray images using ensemble learning with convolutional neural network. *Pattern Anal. Appl.* **2021**, *24*, 1111–1124. [CrossRef]
6. Horry, M.J.; Chakraborty, S.; Paul, M.; Ulhaq, A.; Pradhan, B.; Saha, M.; Shukla, N. X-ray image based COVID-19 detection using pre-trained deep learning models. *IEEE Access* **2020**, *8*, 149808–149824. [CrossRef]
7. Yang, D.; Martinez, C.; Visuña, L.; Khandhar, H.; Bhatt, C.; Carretero, J. Detection and Analysis of COVID-19 in Medical Images Using Deep Learning Techniques. *Sci. Rep.* **2021**, *11*, 19638. [CrossRef] [PubMed]
8. Bansal, S.; Singh, M.; Dubey, R.K.; Panigrahi, B.K. Multi-Objective Genetic Algorithm Based deep learning Model for Automated COVID-19 Detection using Medical Image Data. *J. Med. Biol. Eng.* **2021**, *41*, 678–689. [CrossRef]
9. Murugappan, M.; Bourisly, A.K.; Krishnan, P.T.; Maruthapillai, V.; Muthusamy, H. Artificial Intelligence Based COVID-19 Detection using Medical Imaging Methods: A Review. *Comput. Model. Imaging SARS-CoV-2 COVID-19* **2021**, *2*, 91–107. [CrossRef]
10. Agrawal, T.; Choudhary, P. FocusCovid: Automated COVID-19 detection using deep learning with chest X-ray images. *Evol. Syst.* **2021**, *13*, 519–533. [CrossRef]
11. Zhang, W.; Zhou, T.; Lu, Q.; Wang, X.; Zhu, C.; Sun, H.; Wang, Z.; Lo, S.K.; Wang, F. Dynamic-Fusion-Based Federated Learning for COVID-19 Detection. *IEEE Internet Things J.* **2021**, *8*, 15884–15891. [CrossRef] [PubMed]
12. Abbas, A.; Abdelsamea, M.M.; Gaber, M.M. Classification of COVID-19 in chest X-ray images using DeTraC deep convolutional neural network. *Appl. Intell.* **2021**, *51*, 854–864. [CrossRef] [PubMed]
13. Ghoshal, B.; Tucker, A. Estimating Uncertainty and Interpretability in deep learning for Coronavirus (COVID-19) Detection. *arXiv* **2020**, arXiv:2003.10769.
14. Islam, M.Z.; Islam, M.M.; Asraf, A. A Combined Deep CNN-LSTM Network for the Detection of Novel Coronavirus (COVID-19) Using X-ray Images. *Inform. Med. Unlocked* **2020**, *20*, 100412. [CrossRef] [PubMed]
15. Mahmoudi, S.A.; Stassin, S.; Daho, M.E.; Lessage, X.; Mahmoudi, S. Explainable deep learning for Covid-19 Detection Using Chest X-ray and CT-Scan Images. In *Healthcare Informatics for Fighting COVID-19 and Future Epidemics*; Springer: Berlin/Heidelberg, Germany, 2021; pp. 311–336.
16. Zebin, T.; Rezvy, S. COVID-19 detection and disease progression visualization: Deep learning on chest X-rays for classification and coarse localization. *Appl. Intell. (Dordr. Neth.)* **2021**, *51*, 1010–1021. [CrossRef] [PubMed]
17. Shorfuzzaman, M.; Masud, M.; Alhumyani, H.; Anand, D.; Singh, A. Artificial Neural Network-Based Deep Learning Model for COVID-19 Patient Detection Using X-Ray Chest Images. *J. Healthc. Eng.* **2021**, *2021*, 5513679. [CrossRef] [PubMed]
18. Horry, M.J.; Chakraborty, S.; Paul, M.; Ulhaq, A.; Pradhan, B.; Saha, M.; Shukla, N. COVID-19 Detection Through Transfer Learning Using Multimodal Imaging Data. *IEEE Access* **2021**, *8*, 149808–149824. [CrossRef]
19. Alqudah, A.M.; Qazan, S.; Alqudah, A. Automated Systems for Detection of COVID-19 Using Chest X-ray Images and Lightweight Convolutional Neural Networks. *Sensors* **2020**, *20*, 6985. [CrossRef]
20. Khalifa, N.M.; Taha, M.H.; Hassanien, A.E.; Taha, S.H. The Detection of COVID-19 in CT Medical Images: A deep learning Approach. In *Big Data Analytics and Artificial Intelligence Against COVID-19: Innovation Vision and Approach*; Springer: Berlin/Heidelberg, Germany, 2020; Volume 78, pp. 73–90.
21. Jangam, E.; Barreto, A.A.D.; Annavarapu, C.S.R. Automatic detection of COVID-19 from chest CT scan and chest X-Rays images using deep learning, transfer learning and stacking. *Appl. Intell.* **2021**, *52*, 2243–2259. [CrossRef]

22. Padma, T.; Kumari, C.U. Deep learning Based Chest X-Ray Image as a Diagnostic Tool for COVID-19. In Proceedings of the 2020 International Conference on Smart Electronics and Communication (ICOSEC), Trichy, India, 10–12 September 2020; pp. 589–592.
23. Nawshad, M.A.; Shami, U.A.; Sajid, S.; Fraz, M.M. Attention Based Residual Network for Effective Detection of COVID-19 and Viral Pneumonia. In Proceedings of the 2021 International Conference on Digital Futures and Transformative Technologies (ICoDT2), Islamabad, Pakistan, 20–21 May 2021; pp. 1–7.
24. Ashraf, A.; Malik, A.U.; Khan, Z.H. POSTER: Diagnosis of COVID-19 through Transfer Learning Techniques on CT Scans: A Comparison of deep learning Models. In Proceedings of the 2022 2nd International Conference of Smart Systems and Emerging Technologies (SMARTTECH), Riyadh, Saudi Arabia, 9–11 May 2022.
25. Kumar, N.; Hashmi, A.; Gupta, M.; Kundu, A. Automatic Diagnosis of Covid-19 Related Pneumonia from CXR and CT-Scan Images. *Eng. Technol. Appl. Sci. Res.* **2022**, *12*, 7993–7997. [CrossRef]
26. Kamil, M.Y. A deep learning framework to detect Covid-19 disease via chest X-ray and CT scan images. *Int. J. Electr. Comput. Eng.* **2021**, *11*, 844–850. [CrossRef]
27. Chouat, I.; Echtioui, A.; Khemakhem, R.; Zouch, W.; Ghorbel, M.; Ben Hamida, A. COVID-19 detection in CT and CXR images using deep learning models. *Biogerontology* **2022**, *23*, 65–84. [CrossRef] [PubMed]
28. Patel, S. Classification of COVID-19 from chest X-ray images using a deep convolutional neural network. *Turk. J. Comput. Math. Educ.* **2021**, *12*, 2643–2651.
29. Neha, K.; Joshi, K.P.; Jyothi, N.A.; Kumar, J.V. Preliminary Detection of COVID-19 Using deep learning and Machine Learning Techniques on Radiological Data. *Indian J. Comput. Sci. Eng.* **2021**, *12*, 79–88. [CrossRef]
30. Olcer, D.; Erdaş, Ç.B. A deep learning approach fed by ct scans for diagnosis of COVID-19. *Selcuk. Univ. J. Eng. Sci.* **2020**, *19*, 110–116.
31. Umair, M.; Khan, M.S.; Ahmed, F.; Baothman, F.; Alqahtani, F.; Alian, M.; Ahmad, J. Detection of COVID-19 Using Transfer Learning and Grad-CAM Visualization on Indigenously Collected X-ray Dataset. *Sensors* **2021**, *21*, 5813. [CrossRef]
32. Dandotiya, H. Deep learning-based detection model for coronavirus (COVID-19) using CT and X-ray image data. *Diagnostics* **2021**, *11*, 340. [CrossRef]
33. Shankar, K.; Perumal, E. Automated Detection and Classification of COVID-19 from Chest X-ray Images Using Deep Learning. *J. Comput. Theor. Nanosci.* **2020**, *17*, 5457–5463. [CrossRef]
34. Jain, A.; Ratnoo, S.; Kumar, D.K. Convolutional Neural Network for Covid-19 detection from X-ray images. In Proceedings of the 2021 Fourth International Conference on Computational Intelligence and Communication Technologies (CCICT), Sonepat, India, 3 July 2021; pp. 100–104.
35. Ahmed, D.S.; Allah, H.A.A.A.; Hussain, S.A.-K.; Abbas, I.K. Detection COVID-19 of CT-Scan Image for Hospitalized Iraqi Patients based on Deep Learning. *Webology* **2022**, *19*, 1028–1055. [CrossRef]
36. Yadav, S.S.; Bendre, M.; Vikhe, P.S.; Jadhav, S.M. Analysis of deep machine learning algorithms in COVID-19 disease diagnosis. *arXiv* **2020**, arXiv:PPR:PPR347356.
37. Yadav, S.S.; Sandhu, J.K.; Bendre, M.; Vikhe, P.S.; Kaur, A. A comparison of deep machine learning algorithms in COVID-19 disease diagnosis. *arXiv* **2020**, arXiv:2008.11639.
38. Kumar, S.; Abhishek, K.; Singh, K. COVID-19 Detection from Chest X-Rays and CT Scans using Dilated Convolutional Neural Networks. In Proceedings of the 2021 10th IEEE International Conference on Communication Systems and Network Technologies (CSNT), Bhopal, India, 18–19 June 2021; pp. 369–374.
39. Marappan, R.; Sethumadhavan, G. Solving graph coloring problem for large graphs. *Glob. J. Pure Appl. Math.* **2015**, *11*, 2487–2494.
40. Alakus, T.B.; Turkoglu, I. Comparison of deep learning approaches to predict COVID-19 infection. *Chaos Solitons Fractals* **2020**, *140*, 110120. [CrossRef]
41. Marappan, R.; Sethumadhavan, G. Solution to graph coloring problem using heuristics and recursive backtracking. *Int. J. Appl. Eng. Res.* **2015**, *10*, 25939–25944.
42. Sitaula, C.; Hossain, M.B. Attention-based VGG-16 model for COVID-19 chest X-ray image classification. *Appl. Intell.* **2020**, *51*, 2850–2863. [CrossRef] [PubMed]
43. Pathak, Y.; Shukla, P.; Tiwari, A.; Stalin, S.; Singh, S. Deep Transfer Learning Based Classification Model for COVID-19 Disease. *Irbm* **2022**, *43*, 87–92. [CrossRef]
44. Bhaskaran, S.; Marappan, R. Design and Analysis of an Efficient Machine Learning Based Hybrid Recommendation System with Enhanced Density-Based Spatial Clustering for Digital E-Learning Applications. *Complex Intell. Syst.* **2021**, *1*, 1–17. [CrossRef]
45. Marappan, R.; Sethumadhavan, G. A New Genetic Algorithm for Graph Coloring. In Proceedings of the 2013 Fifth International Conference on Computational Intelligence, Modelling and Simulation, Seoul, Republic of Korea, 24–25 September 2013; pp. 49–54. [CrossRef]
46. Sethumadhavan, G.; Marappan, R. A genetic algorithm for graph coloring using single parent conflict gene crossover and mutation with conflict gene removal procedure. In Proceedings of the 2013 IEEE International Conference on Computational Intelligence and Computing Research, Enathi, India, 26–28 December 2013; pp. 1–6. [CrossRef]
47. Marappan, R.; Sethumadhavan, G. Divide and conquer based genetic method for solving channel allocation. In Proceedings of the 2016 International Conference on Information Communication and Embedded Systems (ICICES), Chennai, India, 25–26 February 2016; pp. 1–5. [CrossRef]

48. Veeramanickam, M.R.M.; Rodriguez, C.; Navarro Depaz, C.; Concha, U.R.; Pandey, B.; Kharat, R.S.; Marappan, R. Machine Learning Based Recommendation System for Web-Search Learning. *Telecom* **2023**, *4*, 118–134. [CrossRef]
49. Marappan, R.; Sethumadhavan, G. Solution to Graph Coloring Using Genetic and Tabu Search Procedures. *Arab. J. Sci. Eng.* **2018**, *43*, 525–542. [CrossRef]
50. Marappan, R.; Sethumadhavan, G. Complexity Analysis and Stochastic Convergence of Some Well-known Evolutionary Operators for Solving Graph Coloring Problem. *Mathematics* **2020**, *8*, 303. [CrossRef]
51. Bhaskaran, S.; Marappan, R.; Santhi, B. Design and Comparative Analysis of New Personalized Recommender Algorithms with Specific Features for Large Scale Datasets. *Mathematics* **2020**, *8*, 1106. [CrossRef]
52. Bhaskaran, S.; Marappan, R.; Santhi, B. Design and Analysis of a Cluster-Based Intelligent Hybrid Recommendation System for E-Learning Applications. *Mathematics* **2021**, *9*, 197. [CrossRef]
53. Hussain Ali, Y.; Chinnaperumal, S.; Marappan, R.; Raju, S.K.; Sadiq, A.T.; Farhan, A.K.; Srinivasan, P. Multi-Layered Non-Local Bayes Model for Lung Cancer Early Diagnosis Prediction with the Internet of Medical Things. *Bioengineering* **2023**, *10*, 138. [CrossRef] [PubMed]
54. Reegu, F.A.; Abas, H.; Jabbari, A.; Akmam, R.; Uddin, M.; Wu, C.-M.; Chen, C.-L.; Khalaf, O.I. Interoperability Requirements for Blockchain-Enabled Electronic Health Records in Healthcare: A Systematic Review and Open Research Challenges. *Secur. Commun. Networks* **2022**, *2022*, 1–11. [CrossRef]
55. Banumathy, D.; Khalaf, O.I.; Romero, C.A.T.; Indra, J.; Sharma, D.K. CAD of BCD from Thermal Mammogram Images Using Machine Learning. *Intell. Autom. Soft Comput.* **2022**, *34*, 595–612. [CrossRef]
56. Tavera, C.A.; Jesús, H.O.; Osamah, I.K.; Diego, F. Saavedra; Theyazn HH Aldhyani. Wearable wireless body area networks for medical applications. *Comput. Math. Methods Med.* **2021**, *1*, 5574376. [CrossRef]
57. Sudhakar, S.; Khalaf, O.I.; Priyadarsini, S.; Sharma, D.K.; Amarendra, K.; Hamad, A.A. Smart healthcare security device on medical IoT using raspberry pi. *Int. J. Reliab. Qual. E-Healthc.* **2022**, *11*, 1–11.
58. Sengan, S.; Khalaf, O.I.; Rao, G.R.K.; Sharma, D.K.; Amarendra, K.; Hamad, A.A. Security-Aware Routing on Wireless Communication for E-Health Records Monitoring Using Machine Learning. *Int. J. Reliab. Qual. E-Healthcare* **2021**, *11*, 1–10. Available online: https://econpapers.repec.org/article/iggjrqeh0/v_3a11_3ay_3a2022_3ai_3a3_3ap_3a1-10.htm (accessed on 15 October 2021). [CrossRef]
59. Marappan, R.; Sethumadhavan, G. Solution to Graph Coloring Problem using Evolutionary Optimization through Symmetry-Breaking Approach. *Int. J. Appl. Eng. Res.* **2015**, *10*, 26573–26580.
60. Marappan, R.; Sethumadhavan, G. Solving Graph Coloring Problem Using Divide and Conquer-Based Turbulent Particle Swarm Optimization. *Arab. J. Sci. Eng.* **2021**, *47*, 9695–9712. [CrossRef]
61. Marappan, R.; Sethumadhavan, G. Solution to graph coloring problem using divide and conquer based genetic method. In Proceedings of the 2016 International Conference on Information Communication and Embedded Systems (ICICES), Chennai, India, 25–26 February 2016; pp. 1–5. [CrossRef]
62. Anand, N.S.; Marappan, R.; Sethumadhavan, G. Performance Analysis of SAR Image Speckle Filters and its Recent Challenges. In Proceedings of the 2018 IEEE International Conference on Computational Intelligence and Computing Research (ICCIC), Madurai, India, 13–15 December 2018; pp. 1–4. [CrossRef]
63. Bhaskaran, S.; Marappan, R. Enhanced personalized recommendation system for machine learning public datasets: Generalized modeling, simulation, significant results and analysis. *Int. J. Inf. Technol.* **2023**, *1*, 1–13. [CrossRef]

bioengineering

Article

In-Silico Lead Druggable Compounds Identification against SARS COVID-19 Main Protease Target from In-House, Chembridge and Zinc Databases by Structure-Based Virtual Screening, Molecular Docking and Molecular Dynamics Simulations

Mehreen Ghufran, Mehran Ullah, Haider Ali Khan, Sabreen Ghufran, Muhammad Ayaz and Muhammad Siddiq et al.

[1] Department of Pathology, Medical Teaching Institution Bacha Khan Medical College (BKMC) Mardan, Mardan 23200, Pakistan
[2] District Medical Officer, Sehat Sahulat Program (SSP), KPK, Mardan 23200, Pakistan
[3] Mardan Medical Complex (MMC) Mardan, Medical Teaching Institution Bacha Khan Medical College (BKMC), Mardan 23200, Pakistan
[4] Department of Biochemistry, Abdul Wali Khan University Mardan, Mardan 23200, Pakistan
[5] Department of Pharmacy, University of Malakand, Chakdara 18000, Pakistan
[6] Department of Pharmacy, Abdul Wali Khan University Mardan, Mardan 23200, Pakistan
[7] Department of Pharmacy, Sarhad University of Science and technology, Peshawar 25000, Pakistan
[8] Shanghai Key Laboratory for Molecular Engineering of Chiral Drugs, School of Pharmacy, Shanghai Jiao Tong University, Shanghai 200240, China
[9] Department of Natural Product Chemistry, School of Pharmacy, Shanghai Jiao Tong University, Shanghai 200240, China
[10] Department of Pharmacy, Faculty of Medicine and Pharmacy, University of Oradea, 410028 Oradea, Romania
* Correspondence: haider.ali@awkum.edu.pk (H.A.K.); shams1327@yahoo.com (S.S.u.H.)

Citation: Mehreen Ghufran, Mehran Ullah, Haider Ali Khan, Sabreen Ghufran, Muhammad Ayaz and Muhammad Siddiq et al. In-Silico Lead Druggable Compounds Identification against SARS COVID-19 Main Protease Target from In-House, Chembridge and Zinc Databases by Structure-Based Virtual Screening, Molecular Docking and Molecular Dynamics Simulations. *Bioengineering* **2023**, *10*, 100. https://doi.org/10.3390/bioengineering10010100

Academic Editors: Sundeep Singh, Roderick Melnik, Esther Pueyo and Ali Zarrabi

Received: 29 December 2022
Revised: 4 January 2023
Accepted: 8 January 2023
Published: 11 January 2023

Abstract: Pharmacological strategies to lower the viral load among patients suffering from severe diseases were researched in great detail during the SARS-CoV-2 outbreak. The viral protease M^{pro} (3CLpro) is necessary for viral replication and is among the main therapeutic targets proposed, thus far. To stop the pandemic from spreading, researchers are working to find more effective M^{pro} inhibitors against SARS-CoV-2. The 33.8 kDa M^{pro} protease of SARS-CoV-2, being a nonhuman homologue, has the possibility of being utilized as a therapeutic target against coronaviruses. To develop drug-like compounds capable of preventing the replication of SARS-main CoV-2's protease (M^{pro}), a computer-aided drug design (CADD) approach is extremely viable. Using MOE, structure-based virtual screening (SBVS) of in-house and commercial databases was carried out using SARS-CoV-2 proteins. The most promising hits obtained during virtual screening (VS) were put through molecular docking with the help of MOE. The virtual screening yielded 3/5 hits (in-house database) and 56/66 hits (commercial databases). Finally, 3/5 hits (in-house database), 3/5 hits (ZINC database), and 2/7 hits (ChemBridge database) were chosen as potent lead compounds using various scaffolds due to their considerable binding affinity with M^{pro} protein. The outcomes of SBVS were then validated using an analysis based on molecular dynamics simulation (MDS). The complexes' stability was tested using MDS and post-MDS. The most promising candidates were found to exhibit a high capacity for fitting into the protein-binding pocket and interacting with the catalytic dyad. At least one of the scaffolds selected will possibly prove useful for future research. However, further scientific confirmation in the form of preclinical and clinical research is required before implementation.

Keywords: main protease (M^{pro}); structure-based virtual screening; ZINC; in-house; ChemBridge database; molecular dynamics simulation

1. Introduction

Since December 2019, after the first outbreak of Corona virus infection reported from Wuhan, China, the disease has devastated life throughout the world and the search for affective therapeutics is underway [1]. The virus is known as SARS-CoV-2 because its RNA genome shares 82 percent of its sequence with the SARS Corona virus [2]. These viruses are related to the betacoronavirus clade B [3,4]. Although the outbreak was initially thought to have originated in Wuhan's Huanan seafood and cattle market, effective human-to-human transmission has caused the number of patients to rise dramatically. As of April 9th, there were more than 1,500,000 diseased individuals with a 5.9% mortality rate. The recent emergence of the coronavirus-2 causing severe acute respiratory illness (SARS-CoV-2) has led to the global pandemic of coronavirus disease 2019 (COVID-19). By April 2021, there had been more than 140 million infections reported, causing more than 3 million fatalities globally. Antiviral medications will likely be essential to manage the anticipated future outbreaks of coronaviruses, despite the promising COVID-19 immunization campaigns. The emergence of SARS-CoV-2 variants for which vaccinations are ineffective suggests that antiviral medications will eventually be needed to enhance immunizations [5]. Similar to the common cold virus, SARS-CoV-2 is expected to continue spreading and provide a significant threat to our society. In this condition, antiviral medications are required to treat infected patients as well as be delivered prophylactically to protect high-risk groups. Since therapeutic medicines that suppress coronavirus replication have the potential to enhance the lives of millions of people throughout the world, their discovery must be prioritized despite the lengthy drug development process. After SARS-CoV-1 (found in 2002) and MERS-CoV (Middle East respiratory illness, 2012), SARS-CoV-2 is the deadliest of the zoonotic coronaviruses that have infected humans [6]. Similar to other coronaviruses, SARS-CoV-2 affects the respiratory system and causes severe pneumonia, which necessitates ventilatory assistance and intensive care, especially in the elderly and immunocompromised patients [3]. Vaccine development has advanced significantly, but supply and timing are currently limiting factors for its effective implementation. Several vaccinations have been developed and licensed for mass immunization [7]. The cost of storing some vaccines at cryogenic temperatures, meanwhile, may be prohibitive in underdeveloped nations. Additionally, a number of changes to the SARS-CoV-2 genome may impact how well vaccines work to fight the virus [8,9]. These results highlight how critical it is to simultaneously develop therapeutic options for SARS-CoV-2 treatment.

SARS-CoV-2 belongs to the beta group of coronaviruses, an RNA virus with only one strand. It has structural proteins, such as spike-like protein S and lipid membranes, as well as M protein (membrane), N protein (nucleo-capsid), and envelope (E) protein that give it an envelope appearance. The S spike protein binds to the angiotensin-converting enzyme 2 (ACE2) receptor on mammalian lung cells, allowing the virus's RNA genetic material to be released into the host cells [10]. Four nonstructural proteins are found in the virus: papain-like (PLpro) and CoV main proteases [M^{pro}; also known as 3CLpro] [11], RNA polymerase, and helicase [12]. The virus's transcription and replication are aided by both proteases (PLpro and 3CLpro). The replicase genes encode two polyproteins important for effective viral replication and transcription [13]. A significant proteolytic process liberates the functional polypeptides from these two polyproteins (pp1a and pp1ab). Proteolysis is mostly performed by a papain-like protease (PLpro), which cuts proteins in three places, and a 33.8 kDa main protease (Mpro), also called a 3C-like protease, which cuts proteins in 11 sites, making nonstructural proteins in the process (NSPs). As no host protease recognizes the M^{pro} recognition sequence, to develop drugs against SARS-CoV-2 infections, this enzyme is a prime target for the development of inhibitors. The Mpro and PLpro enzymes, which digest viral polyproteins produced by the host cell translational machinery, build a functionally active viral replication complex and package it into host cells during viral replication [14]. Three domains make up the 3CLpro monomer, with the active site (Cys 145 and His 41) located between domains I and II. A larger pocket is found in the gap between the third domain and the protein structure because of the long loop connecting it to the rest [15]. Additionally, 3CLpro is the main protease of the virus, and it helps in the replication of the virus, making it a valuable antiviral treatment target. M^{pro} is a crucial target for antiviral medication because the human genome lacks a homologue of it. Although there is no homologue of M^{pro} in the human genome, it is known as 3CLpro (3-chymotrypsin-like protease) and aids in viral replication, making it a significant target for antiviral treatments. Protease inhibitors efficiently stop coronavirus replication and proliferation by obstructing the post-translational processing of essential viral polypeptides [16]. Pfizer's PF-07321332 is an oral antiviral compound that is designed to stop SARS-CoV-2 Mpro from modifying the active site Cys145 with its nitrile warhead. It is thought to be a good antiviral candidate and is currently being tested (NCT04756531, NCT04909853, NCT05011513, and ClinicalTrials.gov (accessed on 23 March 2021)). The oral antiviral PAXLOVIDTM, which is a combination of PF-07321332 and the HIV drug ritonavir, which slows down the breakdown of PF-07321332, was found to reduce the risk of hospitalization or death by 89% compared to a placebo in nonhospitalized high-risk adults with COVID-19. In December 2021, the FDA gave Pfizer's Paxlovid an emergency use authorization to treat mild-to-moderate COVID-19 in adults and children older than 12 who are at least 12 years old (www.fda.gov) [17]. Crystallization of the main protease of SARS-CoV-2 (PDB ID: 6LU7) has been accomplished by Liu et al., which provides an opportunity to combat the disease by identifying it as a potential therapeutic target. When opposed to methods based on trial and error that involve experimental research, the use of the in silico method for the screening of prospective therapeutic compounds has been demonstrated to be both time and cost efficient. The in silico method of molecular docking has the capacity to screen and find potentially useful therapeutic compounds from large and huge compound databases. At the moment, several molecular docking studies are being conducted against SARS-CoV-2 receptors with drug-like compounds (ChemBridge, ZINC, and in-house databases). Furthermore, the majority of these docking investigations used quantitative structure-activity relationships (QSARs) modeling, similarity searches, and structure-based drug design (SBDD) [18].

Drug repurposing is one method for speeding up the normal drug development and drug discovery process. This will help to clarify the effectiveness of novel therapeutic uses for substances whose efficacy and safety have already been established. The importance of potential therapeutic compounds that function as efficient antivirals in controlling the pandemic is highlighted by the increasing and faster spread of SARS-CoV-2 as well as the appearance of novel variants. In the formation of anticoronavirus therapeutics, $M^{pro'}$s catalytic activity may be hindered. In light of this background, the current research looked for a few phytocomponents that could suppress the M^{pro} protein. Phytochemicals are used as active drugs in drug development. In the fight against viral diseases, phytocompounds derived from a variety of medicinal plants may enhance immune function and combat pathogens. Phytochemicals and their derivatives have been the focus of numerous studies due to their antiviral activities and mechanisms of action, which have been demonstrated to be crucial in the treatment of viral diseases [19]. Over the past decade, computer-aided drug discovery (CADD) approaches have emerged as a crucial component of the drug development process, having been used to identify protein inhibitors and investigate interactions between proteins and drugs and proteins themselves [20]. Despite the time and money required to develop a candidate drug into an approved drug, computational methods, such as virtual screening, docking, molecular dynamics (MD) simulations [21], and binding free energy evaluation, can be used to identify promising drug candidates from compound libraries. The purpose of this research was to find potential anti-SARS-CoV-2 treatments by utilizing a multipronged strategy that included both molecular docking and virtual screening techniques. The information that is obtained from the screening will be helpful in the investigation of new inhibitors of the SARS-CoV-2 M^{pro} target that have the potential to be both effective and selective.

2. Materials and Methods

2.1. Protein Preparation

A worldwide database called the Protein Data Bank (PDB) was used to find the three-dimensional structure of main protease (M^{pro}) with co crystallized ligand structure, which causes severe acute respiratory syndrome (PDB Id: 6LU7) (accessed on 23 March 2021) (https://www.rcsb.org). The main protease is composed of two chains, such as A and C. Chain A is the protease, and chain C is the N-[(5methylisoxazol-3yl) carbonyl] alanyl-L-valyl-N-1-(1R,2Z) (1R,2Z)-4-(benzyloxy) (benzyloxy) [(3R)-2-oxopyprolidin-3-3yl]-4-oxo-1-1-[(3R)-2-oxopyprolidin-3-3yl] but-2-enyl methyl)-L-lucinamide (N3 inhibitor). A peptide inhibitor (N3 inhibitor) was in a complex with the protein. Water molecules, inhibitors, and other heteroatoms were removed from the protein structure. All the atoms' Amber14:EHT (Amber ff14SB and ETH combined) forcefields were used to refine the protein structure. The missing hydrogens were added to the amino acids. The forcefield parameters, missing atom types, bond stretch parameters, missing angles, and missing van der Waals parameters were added to all atoms. Restrained electrostatic potential atomic partial charges (RESP) and AM1-BCC [22] charges were used for the protein-ligand complex, the RESP, and AM1-BCC charges were used for leap protein preparation and for preparation of inhibitors, respectively. The 3D protonation was used to incorporate hydrogen atoms into the protein structure, followed by minimization of energy with the MOE (molecular operating environment) program's default parameters [23].

2.2. Preparation of Databases

Small molecules used in virtual screening can be found in commercial databases, such as ZINC and ChemBridge. ZINC had millions of compounds, compared to the ChemBridge database that contain 168423 ligands/compounds in ChemBridge [24]. A Tanimoto cut-off level of 60% was used to screen the ZINC database, which resulted in the production of a library containing 11,193 drug-like molecules [25]. An in-house database containing compounds isolated or synthesized by our collaborators, with a focus on natural products and structural analogues. The in-house database contains over 1600 compounds, representing a wide range of structural diversity across a wide variety of core scaffolds and substitution patterns. Three-dimensional protonation (MMFF94x force field) and energy minimization (constrained minimization of 0.01 Kcal/Å^2 was performed to optimize the ligand structures) using MOE were performed on all of the compounds in the in-house database. Anti-COVID-19 lead compounds can be found using structure-based virtual screening, which scans both in-house and commercial databases.

2.3. Structure-Based Virtual Screening

The drug target (receptor) and ligands' 3D structures in the database are necessary for the structure-based virtual screening (SBVS) methods. In order to find new potential inhibitors, we employed molecular docking approach to assess the binding modes of drug target proteins and the ligands, called the structure-based drug design (SBDD). This helps to predict the improved and healthy interactions that will take place between the target receptor and the drug. Using the MOE, the SBVS was used to screen ZINC, in-house, and ChemBridge databases. Additionally, using both revised-Lipinski's rule of 5 as well as Lipinski's rule of 5, the number of screened results was significantly reduced. New lead hits discovered through screening a compound database of thousands of compounds are illustrated by the property of "drug-likeness" [26].

2.4. Molecular Docking

The MOE docking software was utilized to conduct the molecular docking studies [27]. The retrieved compounds were docked with M^{pro} to further evaluate these drug-like compounds. The 3D protonation of the target receptor was followed by energy minimization using the MOE software 2019's default parameters to achieve the best possible outcome. To improve the result, all of the compounds were docked into the $M^{pro'}$ binding pocket. MOE was used to dock the retrieved hits against the M^{pro} drug target. Ten conformations were generated for each hit, with the top-ranked conformations of each inhibitor being used for advanced research. The docking analysis was scrutinized more closely, with docking scores and protein/hit interactions playing a larger role. In addition, the results of molecular docking were validated using MDS.

2.5. Molecular Dynamics Simulation (MDS)

A molecular dynamic (MD) simulation was carried out to investigate the dynamic behavior of proteins upon inhibitor binding at the atomic level. The docked conformations of the selected hits within the active pocket of M^{pro} were subjected to MD simulations. A detailed MDS analysis was carried out using the Amber14 package and the ff14SB force field [28]. To evaluate the stability of the previously retrieved compounds at the active sites of M^{pro} and M^{pro}/N3 complex, MDS was used. Tleap, a preparatory program, was used to build and solve the complexes. The solvated octahedral box was used in this experiment. After solvating each system in an octahedral box using the TIP3P water model with 15 Å, the systems were neutralized by adding counterions (either Na+ or Cl−). Each neutralized system's energy was reduced as much as possible through two steps of energy minimization in order to achieve the goal of relaxing all of the systems. These steps were steepest descent minimization and conjugate gradient minimization. At 50 ps, the minimized complexes were heated to 300 K. Then, using a two-step process, each system was brought into equilibrium at a constant 1 atom and 300 K. First, we used a weak restraint to equilibrate the density at 50 ps. Second, we equilibrated the system without any constraints for 1 ns. After that, the production step was run for 150 ns. To keep the temperature stable, the Langevin thermostat was activated [29] and Berendsen barostat was used to monitor the system pressure. For the calculation of long-range electrostatic interaction, we used the AMBER18 Particle Mesh Ewald (PME) algorithm. A cut-off distance of 10 Å was used for long-range electrostatic interactions and van der Waals interactions. The covalent bonds were refined using the AMBER18 SHAKE algorithm [30]. The GPU version (PMEMD.cuda) [31] of AMBER18 was used to run MD simulations on four complexes with M^{pro}/N3 complexes. The AMBER18 CPPRTAJ module was used to analyze the MD trajectories. The interface analysis and graphical representation were carried out using MOE2019 software, PyMol v1.7, and Origin Pro Lab v2018.

2.6. Assessment of Binding Free Energy

For the calculation of binding free energy (BFE), trajectories generated by molecular dynamics simulations using the MMPBSA.py script were used [32]. Numerous studies have employed this method to evaluate the binding free energies of P-P (protein-protein), protein-ligand, and nucleic acid-protein complexes. The total binding free energy (Gbind) was calculated with the help of the following Equation:

$$\Delta \text{Gbind} = \Delta \text{Gcomplex} - [\Delta \text{Greceptor} + \Delta \text{Gligand}]$$

For each of the energy terms, for example, polar (Gpol), van der Waal forces (GvdW), electrostatic energy (Gele), no-polar interactions (Gnpol),Gbond showed the angle of bond and their dihedral energy, TS represents the absolute temperature (T) and entropy (S), the equation below was used to better understand how they contribute to the total energy (G).

$$\text{G} = \text{Gele} + \text{Gbond} + \text{GvdW} + \text{Gnpol} + \text{Gpol} - \text{TS}$$

The molecular mechanics generalized born surface area (MM-GBSA) method was used to calculate the binding free energies of the retrieved compounds/complexes and M^{pro}/N3. Since the MM-GBSA is a BFE index, the lower the value, the stronger the bond. The binding free energy of the retrieved and reference complexes was calculated in Amber 18 using the python script MMGBSA.py. The decrease in potential energy over 150 ns revealed that the system is stable in the case of complexes. The various conformations obtained over a 150 ns simulation period are examined. The MMGBSA.py script was used to calculate the BFE between M^{pro} and the retrieved hits, as well as the reference drug (N3, peptide inhibitor) [33]. In this study, the MMGBSA scripts from AMBER and AMBER Tools were used to carry out various steps required to assess the BFE of the protein-ligand complex via MMGBSA methods. By taking 15,000 snapshots over a 150-ns trajectory, the BFE was calculated.

3. Results and Discussion

Despite the encouraging vaccination programs against COVID-19, the use of antivirals is likely going to be necessary in order to contain the unpredictable outbreaks of coronaviruses that will occur in the future. Vaccines have already been developed, and there are SARS-CoV-2 variants that are resistant to them, which is clear evidence that antivirals will eventually need to be used in addition to vaccines [5]. SARS-CoV-2 must, therefore, be controlled using an antiviral drug that is both affordable and effective. To discover new drugs, it is helpful to see if existing drugs or drugs with similar properties are effective in the treatment of viral infections. The traditional methods of drug discovery take a long time and are inefficient [34]. According to an in silico study, the N3 inhibitor blocks the active catalytic site of HCov-NL-63, preventing its biological function [35]. The current project aimed at performing structure-based virtual database screening, molecular docking, and drug-likeness evaluations of potential compounds. With the help of this methodology, potent drug candidates were found to bind closely to the M^{pro} of the SARS-CoV-2 catalytic site and limit its proteolytic activity. Through simulation approaches, potential anti-SARS-CoV-2 M^{pro} can be identified by using the structure of the COVID-19 virus M^{pro} in complex with N3. SARS-CoV-2 M^{pro} candidate inhibitors were virtually tested using three databases (ChemBridge, ZINC, and in-house).

3.1. *Structure-Based Virtual Screening*

One of the most useful and effective in silico techniques for the drug design process is the structure-based virtual screening (SBVS) method. SBVS makes an effort to anticipate the interaction mode that will result in the formation of a stable complex between two molecules. It does this by employing scoring functions, which measure the force of noncovalent interactions that occur between a receptor and an inhibitor. Therefore, scoring functions are the primary factors that determine whether or not SBVS software is successful. It is possible to get different results from different software programs, even when utilizing the same input, because these programs all employ different algorithms to perform SBVS, which means that there are many distinct software programs that are used to perform SBVS. In SBVS, the three-dimensional structure of the target protein is already known, and the purpose of the process is to choose ligands from a candidate database in such a way that they will have a greater affinity for the three-dimensional structure of the target. Molecular docking is a computer approach that can be used to perform VS. During this procedure, ligands are moved around in three-dimensional space in an effort to locate a target and ligand combination that maximizes the scoring function. The ligands in the database are ranked according to the highest score they received, and the highest-scoring ones are the ones that can be explored further. For instance, one could look at the mode and kind of interaction that takes place [36]. Through in silico screening, lead inhibitors for the COVID-19 virus M^{pro} can be found by using the structure of the COVID-19 virus M^{pro} in complex with N3. To do this, MOE was used to dock possible binding compounds from the ChemBridge, ZINC, and in-house databases. All the compounds from the different databases (ChemBridge, ZINC, and in-house) were used for virtual screening using MOE software with the M^{pro}. Finding hits/compounds that were chemically and structurally comparable required the SBVS of 3D databases, including the in-house, ChemBridge, and ZINC databases. To find prospective, potential, and new inhibitors, VS was performed on the in-house database, which included 1600 ligands, as well as the ZINC and ChemBridge databases. Five hits from the in-house database, 50 hits from ChemBridge, and 16 hits from the ZINC database were all reported by the SBVS. To verify the draggability of the hits, the retrieved results from the commercial database were also subjected to Lipinski's rule of five, while the results from the in-house database were treated with a modified version of Lipinski's rule of (5) five. According to Lipinski's rule of (five) 5, druggable molecules must have the log S-score $\leq$5, a MW <500 Dalton, an HBA of <10, a log p-score of $\leq$5, and an HBD of <5. These factors are all indicators of H-bond donors. Molecules which did not fit these requirements; their absorptions would be unsatisfactory [37]. On the other hand, the "modified Lipinski's rule of five" suggests that molecules having MW >500 logP, HBD greater than 5, and HBA greater than 10 exhibit good absorption. After running the retrieved hits via both Lipinski's rules [38], it was discovered that, retrieved hits from 49/50 ChemBridge database, 7/16 ZINC database, and 3/5 (in-house database) were following both rules effectively. Subsequently, the 3/5 (in-house) and 56/66 (commercial databases) retrieved hits were further decreased for further analysis using a molecular docking strategy.

3.2. Molecular Docking

By docking all the hits from the 3/5 (in-house database) and 56/66 (commercial databases) using the M^{pro} binding pockets via MOE, results in this study were further reduced and refined. In comparison to the standard drug (N3), the docking scores of our refined hits were considerably good (Table 1). For each ligand in the retrieved findings, ten distinct conformations were generated, and all hits having high conformations were sorted and kept in a database file for subsequent analysis. We observed that 3/5 in-house hits, 7/49 ChemBridge database hits, and 5/7 ZINC database hits were best on the basis of docking scores and, thus, were selected for further analysis. Furthermore, the top-ranked conformations of 3/5 hits (in-house), 3/5 hits (ZINC database), and 2/7 hits (ChemBridge database) were well accommodated inside the active site of the M^{pro} drug target and were implicated in several interactions at the active sites of the target protein. From three different databases (ChemBridge, ZINC, and in-house database), docking calculations revealed eight chemically varied molecules having a better binding affinity towards SARS-CoV-2 M^{pro} as compared to N3. Docking studies indicated that the M^{pro} drug target exhibits better docking scores and considerable polar contacts with the hits due to the presence of electronegative capabilities. Table 1 shows the results of molecular docking. The retrieved potential antiviral (anti-COVID19) hits were found to be well-fitted inside the M^{pro} drug target (Figure 1). The compounds reported in Table 1 were ChemBridge, ZINC, and in-house database compounds, and these retrieved compounds were piperidine derivative, tetrahydrothiophene derivative, triazin analog, pyridazine derivative, triazolo pyridine and quinoxaline derivatives, respectively. These compounds have a role in the inhibition of the M^{Pro} target [39].

Table 1. The finalized lead hit compounds' 2D structures and docking values.

S. NO	Compound Names	Structures	Docking Scores
1	ZINC08535852	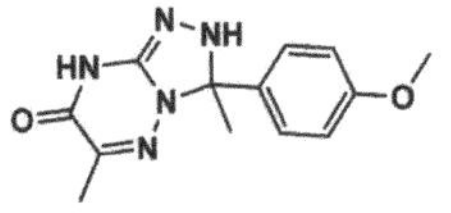 3-(4-methoxyphenyl)-3,6-dimethyl-2,8-dihydro-[1,2,4]triazolo[4,3-b][1,2,4]triazin-7(3H)-one	−41.3801

Table 1. *Cont.*

S. NO	Compound Names	Structures	Docking Scores
2	ZINC44928678	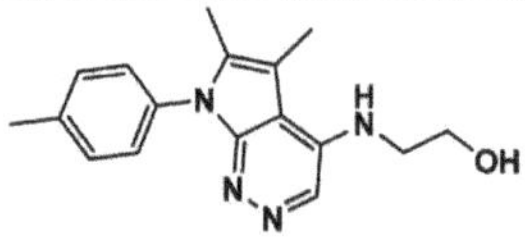 2-((5,6-dimethyl-7-(p-tolyl)-7H-pyrrolo[2,3-c] pyridazin-4-yl)amino)ethan-1-old	−41.0291
3	ZINC72171104	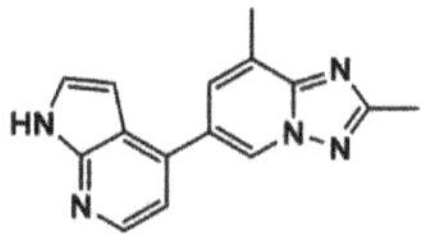 2,8-dimethyl-6-(1H-pyrrolo[2,3-b]pyridin-4-yl)-[1,2,4]triazolo[1,5-a]pyridine	−39.5487
4	12-quinoxaline derivative	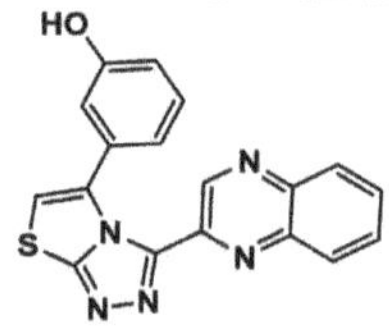 3-(3-(quinoxalin-2-yl)thiazolo[2,3-c] [1,2,4]triazol-5-yl)phenol	−38.7102
5	ChemBridge63310525	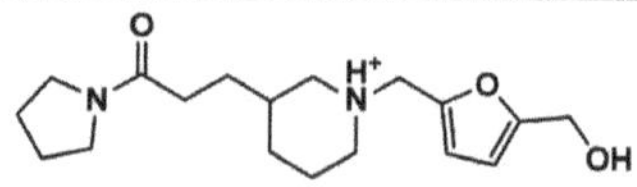 1-((5-(hydroxymethyl)furan-2-yl)methyl)-3-(3-oxo-3-(pyrrol propyl)piperidin-1-ium	−38.0478
6	18-quinoxaline derivative	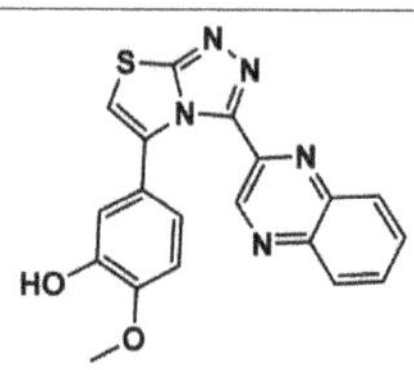 2-methoxy-5-(3-(quinoxalin-2-yl)thiazolo [2,3-c][1,2,4]triazol-5-yl)phenol	−37.5300
7	ChemBridge53208972	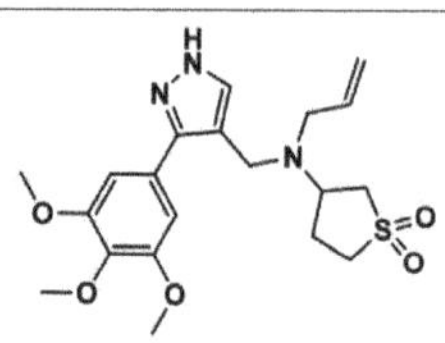 3-(allyl((3-(3,4,5-trimethoxyphenyl)-1H-pyrazol-4-yl)methyl)amino)tetrahydrothiophene 1,1-dioxide	−35.4302

Table 1. *Cont.*

S. NO	Compound Names	Structures	Docking Scores
8	25-quinoxaline derivative	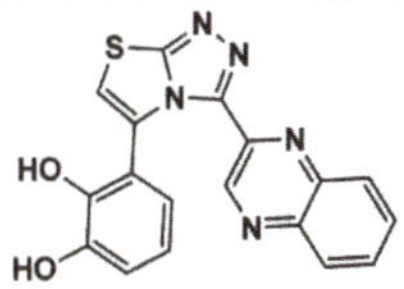 3-(3-(quinoxalin-2-yl)thiazolo[2,3-c][1,2,4]triazol-5-yl) benzene-1,2-diol	−34.3177
Reference	N3	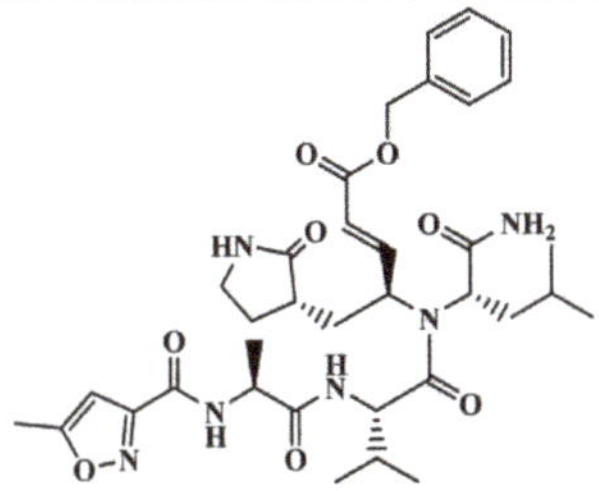 N-[(5-Methyl-3-isoxazolyl)carbonyl]-L-alanyl-L-valyl-N-[(1S,2E)-4-oxo-1-[[(3S)-2-oxo-3-pyrrolidinyl]methyl]-4-(phenyl 2-buten-1-yl]-L-leucinamide	−29.5841

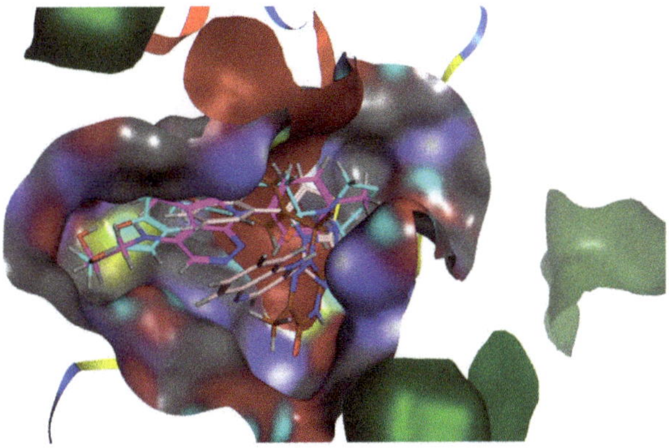

Figure 1. The M^{pro} protein's molecular surface representation with an overlay of all retrieved active hits in the binding pocket. The ZINC44928678, ZINC08535852, 12-quinoxaline derivative, and ChemBridge63310525 active ligands were represented by purple, dark brown, pink, and cyan colors, respectively.

3.3. Analyses of the Binding Interactions of Finally Selected Drug-like Compounds

It is widely recognized that molecular docking provides essential guidelines for the design and discovery of novel drugs. The S-score quantifies the strength of the receptor-ligand interactions. The compounds can be chosen as good drug compounds based on their docking score (S-score). According to the docking study, all finalized hits exhibited favorable contacts with the residues of the M^{pro} target's binding site when compared to the positive control (Table 2).

Table 2. Details of binding interactions against COVID-19 M^{pro}.

Compounds IDs	Ligand	Receptor			Interaction	Distance	E (kcal/mol)
ZINC08535852	C1 1	SG	CYS	44	H-donor	3.91	−0.2
	N1 3	SG	CYS	44	H-donor	3.34	−0.6
	N4 7	O	PRO	52	H-donor	3.09	−1.1
	C6 9	O	ASN	51	H-donor	3.22	−0.1
	NH 11	O	Tyr	54	H-donor	3.69	−0.1
	N3 6	NH1	ARG	188	H-acceptor	3.01	−0.7
	O2 19	CA	ARG	188	H-acceptor	3.85	−0.1
ZINC44928678	C1 1	OG1	THR	24	H-donor	3.65	−0.1
	C1 1	O	THR	24	H-donor	3.65	−0.1
	C6 6	SD	MET	49	H-donor	3.82	−0.2
	C14 18	SD	MET	165	H-donor	3.58	−0.1
	C16 21	SG	CYS	145	H-donor	3.88	−0.1
	C17 22	SG	CYS	145	H-donor	4.1	−0.3
	O1 20	NE2	HIS	41	H-acceptor	2.98	−0.9
ZINC72171104	N2 8	SG	CYS	145	H-donor	3.33	−2.2
	N5 20	O	THR	190	H-donor	2.83	−4.7
	5-ring	CA	ASN	142	pi-H	4.1	−0.4
	5-ring	CA	MET	165	pi-H	3.51	−0.3
	6-ring	CB	MET	165	pi-H	3.59	−0.5
	6-ring	CD	PRO	168	pi-H	4.88	−0.3
	5-ring	CD	PRO	168	pi-H	4.43	−0.3
	5-ring	CA	GLN	189	pi-H	3.64	−1
	6-ring	CG	GLN	189	pi-H	4.21	−0.7
12-quinoxaline derivative	C3 3	ND1	HIE	172	H-donor	3.86	−0.2
	C24 24	O	HIE	164	H-donor	3.43	−0.3
	O25 25	OE1	GLU	166	H-donor	2.53	−4.7
	O25 25	NH	GLU	166	H-acceptor	2.4	−4.8
	N12 12	N	CYS	145	H-acceptor	3.12	−0.8
	N11 11	N	CYS	145	H-acceptor	3.12	−0.8
	N4 4	CA	MET	165	H-acceptor	3.25	−0.1
	O25 25	N	GLU	166	H-acceptor	2.86	−1.3
18-quinoxaline derivative	C3 3	SG	CYS	44	H-donor	2.97	−0.2
	O27 27	NH	ASN	142	H-donor	2.8	−0.6
	O27 27	O	GLY	143	H-acceptor	2.5	−0.9
	C24 24	NH	GLY	143	H-acceptor	1.6	−1.8
	C26 28	SG	CYS	145	H-donor	2.39	−0.1
	6-ring	CG	MET	49	pi-H	4.73	−0.1
	6-ring	CG	MET	49	pi-H	3.86	−1
	C3 3	CA	MET	49	pi-H	4.23	−0.8
25-quinoxaline derivative	C3 3	SD	MET	49	H-donor	2.84	−0.2
	O26 26	O	GLU	166	H-donor	2.7	−2.3
	O25 25	NH	GLU	166	H-donor	2.8	−2.1
	N13 13	CB	THR	190	H-acceptor	2.58	−0.2
	N7 7	CA	GLN	189	H-donor	3.36	−0.3
	N10 10	NH	ARG	188	H-acceptor	3.31	−0.1
	5-ring	N	THR	190	pi-H	3.67	−0.3
ChemBridge63310525	C9 9	O	ARG	188	H-donor	2.93	−0.5
	C11 11	NH	ARG	188	H-donor	2.5	−0.8
	C8 8	SD	CYS	145	H-donor	3.59	−0.1
	O19 19	O	GLU	166	H-donor	3.44	−0.4
	C14 14	O	MET	49	H-donor	3.13	−0.2
	C22 22	OH	THR	190	H-acceptor	2.53	−0.1
	C20 14	C	MET	49	H-donor	3.13	−0.2

Table 2. *Cont.*

Compounds IDs	Ligand	Receptor			Interaction	Distance	E (kcal/mol)
ChemBridge53208972	C15 22	O	HIE	41	H-donor	3.75	−0.1
	C18 28	O	HIP	164	H-donor	3.3	−0.2
	C19 29	SD	MET	49	H-donor	4.48	−0.1
	C18 28	5-ring	HIE	41	H-pi	3.58	−0.1
	6-ring	CA	ARG	188	pi-H	4.02	−0.4
	6-ring	CD	ARG	188	pi-H	4.42	−0.1
	5-ring	CD	ARG	188	pi-H	4.52	−0.1
	6-ring	N	GLN	189	pi-H	4.76	−0.3
N3 (reference ligand)	N 13	O	THR	190	H-donor	2.85	−2.6
	N 23	O	GLU	166	H-donor	2.83	−4.8
	N 39	OE1	GLN	189	H-donor	2.93	−3.3
	O 85	N	GLY	143	H-acceptor	2.80	−1.0
	CD1 50	5-ring	HIS	41	H-pi	4.08	−0.5

Out of all of the hits identified by using the SBVS against the ZINC, ChemBridge, and in-house databases, ZINC08535852 (ZINC database), with a docking score of −41.3801, was the most active ligand of the ZINC database and demonstrated strong interactions with the active site residues of the M^{pro} of SARS-CoV-2. Figure 2 depicts the docking conformations of a selected compound during the docking process. When used in conjunction with the 13 amino acid residues in the active site of the main protease, the ligand formed seven hydrogen bonds within three degrees of freedom with five amino acid residues, namely Cys 44, Asn 51, Pro 52, Tyr 54, and Arn 188 residues (Figure 2). When there are more hydrogen bonds, the binding efficiency and inhibition are both increased as a result [40]. Six-methyl-3,4-dihydro-1,2,4-triazin-5(2H)-one moiety: The sulfur atom of Cys 44 amino acid formed two hydrogen donor interactions with the nitrogen and methyl groups of the triazin-5(2H)-one moiety. During the formation of the H-donor interaction between the carbon atom of the 3-methoxycyclohex-1-ene moiety and the carbon atom of Asn 51, an H-donor interaction is formed. The nitrogen atoms of the 5-methyl-4,5-dihydro-1,2,4-triazole moiety interacted with the carbonyl oxygen of the Pro 52 and the amino group of the Arg 188 through H-donor and H-acceptor bonds formed by the amino group of the Arg 188. It has also been discovered that the nitrogen and oxygen atoms of the 6-methyl-3,4-dihydro-1,2,4-triazin-5(2H)-one moiety of the compound form H-bonds with the active residues Tyr 54 and Arg 188 of the main protease, confirming previous findings. Further, the hydrophobic interactions between His 41 and Met 49 are demonstrated (Figure 2).

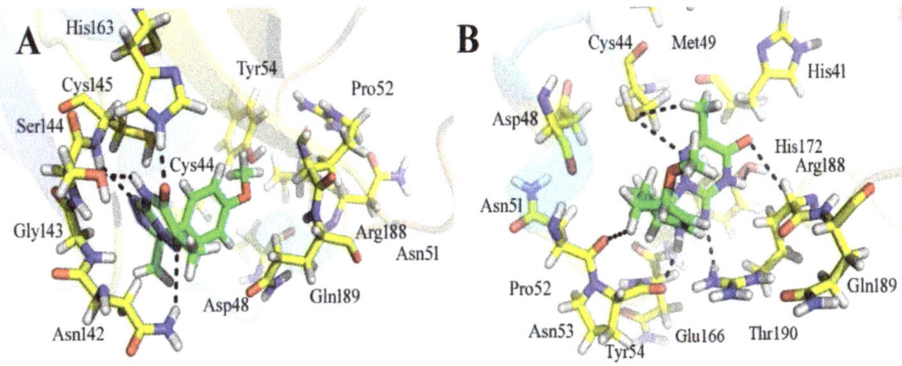

Figure 2. The binding mechanism of the ZINC08535852 ZINC database ligand within the active site of the M^{pro} protein (**A**) before MDS and (**B**) after MDS.

The compound ZINC44928678 (ZINC database) had a good docking score of -41.0291 and good interactions with the target protein's active residues. This compound formed seven polar interactions with the active site residues (Thr 24, His 41, Cys 44, Met 49, Cys 145, and Met 165) according to its binding mode (Figure 3). The carbonyl oxygen and carbon atoms of the 2-(pyridazin-4-ylamino) acetaldehyde moiety of the compound interacted with His 41 and Met 165 residues. The toluene moiety's methyl group formed two H-donor bonds with the Thr 24 residue, while Met 49 showed H-donor interaction with the toluene moiety's carbon atom. The dimethyl groups of the 5,6-dimethyl-7H-pyrrolo [2,3-c] pyridazine moiety of the ligand form two H-donor linkages with Cys 145.

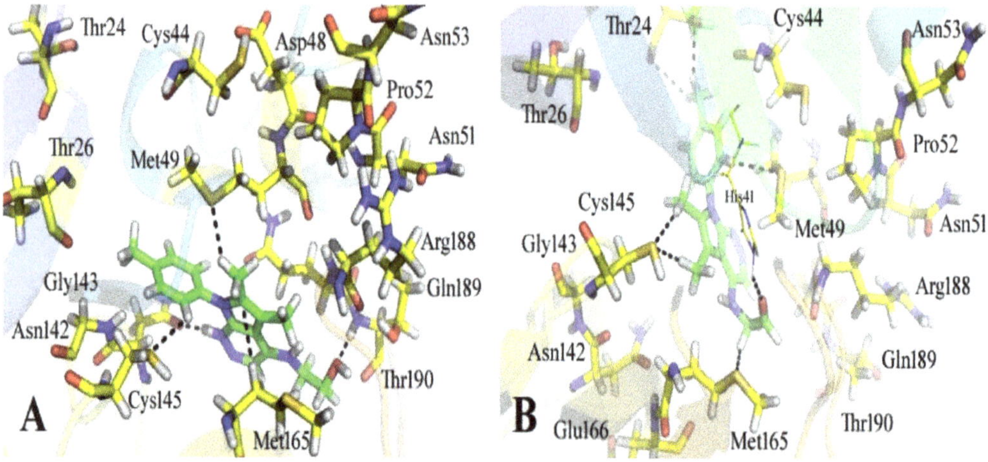

Figure 3. The binding style of the ZINC44928678 ZINC database hit at the M^{pro} protein's active site (**A**) before and (**B**) after MDS.

The 12-quinoxaline derivative from the in-house database had the highest activity among the compounds, with a docking score of −38.7102. It was predicted, based on the docking conformations of the ligand 12-quinoxaline derivative, that the 12-quinoxaline derivative would form hydrophilic and hydrophobic contacts with the active residues of the M^{pro} protein. These active residues include Cys 44, Met 49, Asn 51, Pro 52, Asn 53, Leu 141, Asn 142, Gly 143, Cys 145, His 164, Met 165, Glu 166, His 172, Arg 188, Gln 189, and Thr 190 of M^{pro} protein. It was revealed that the compound formed eight polar interactions with the active residues of the receptor. The H-bond was observed between -NH group of the Asn 142 and nitrogen atom of the thiazolo [2,3-c][1,2,4]triazole moiety of the inhibitor. Cys 145 residue forms two H-acceptor bonds with the nitrogen atoms of the thiazolo [2,3-c][1,2,4]triazole moiety. The carbonyl oxygen atom and –NH group of the Glu 166 residue are connected to the OH group of the phenol moiety of the hit via H-bonds. His 164, Met 165, and His 172 residues were seen to form three H-bonds with the 1,4-dihydroquinoxaline moiety of the compound, as shown in Figure 4.

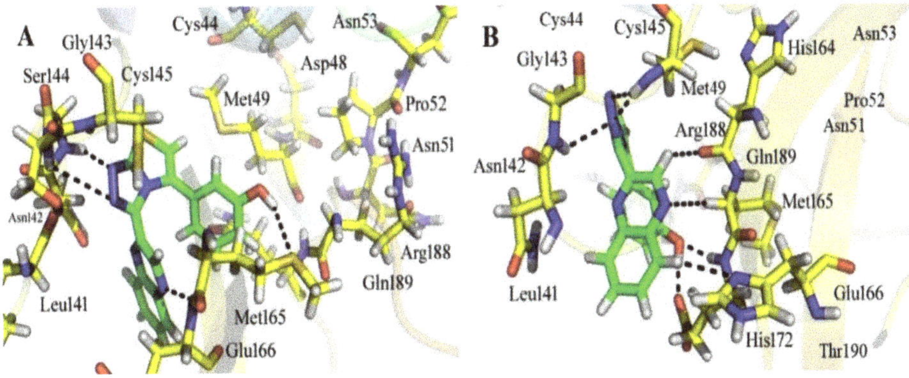

Figure 4. The binding mode of the 12-quinoxaline derivative of the in-house database within the active site of M^{pro} protein (**A**) before and (**B**) after MDS.

ChemBridge63310525 retrieved active hits from the ChemBridge database, forms seven hydrogen bonds, and has a high docking score (−38.0478). On the binding site, ChemBridge63310525 interacts with five important residues and comes out on top. Met 49, Cys145, Glu 166, Gln 189, and Thr 190 form hydrogen bonds with the moieties of the ChemBridge63310525. The hit's furan-2-ylmethanol moiety binds to Cys 145 and Glu 166 via polar bonds. By hydrogen bonding, the 1-(pyrrolidin-1-yl)butan-1-one moiety interacts with Gln 189 and Thr 190. Met 49 forms H-bonds with the piperidin-1-ium moiety of the ligand. Asn 51, Glu166, and Pro 52 interact hydrophobically with the rest of the structure (Figure 5). The compounds that interacted more strongly than the reference inhibitor with the M^{pro} of SARSCoV-2 are listed in Table 1.

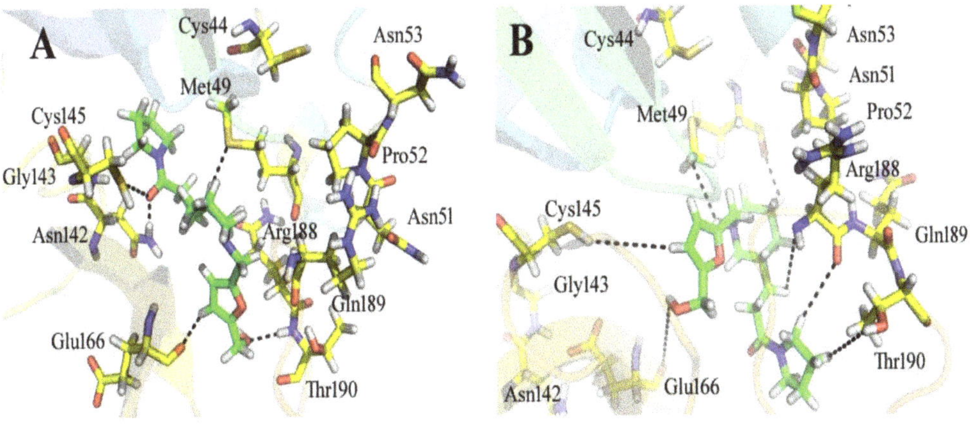

Figure 5. The active site of the M^{pro} protein (**A**) before and (**B**) after MDS, showing the binding mechanism of the obtained inhibitor ChemBridge63310525 from the ChemBridge database.

The retrieved hit compounds are powerful and polarizable due to the electronic cloud density of benzene rings (delocalized electrons), the presence of OH and NH, as well as electron repulsion groups (-nitro). All hits from the study displayed good M^{pro} interactions and docking scores. All of the lead compounds' docking scores and 2D structures are given in Table 1.

3.4. MD Simulation Analysis of the Final Lead Hit/M^pro^ Complexes

The AMBER 18 software was used to perform a 150-ns MDS on the retrieved hit compound/M^{pro} complexes in order to determine their well-stabilized and equilibrated structures. The AMBER 18 software was used to perform MDS on the topmost active retrieved compound/M^{pro} systems, two inhibitors from the ZINC database (ZINC08535852, ZINC44928678), one hit compound from the in-house database (12-quinoxaline derivative), and one compound from the ChemBridge database (ChemBridge63310525) in complex with the M^{pro} drug target, as well as the reference-ligand (N3) M^{pro} complex. Calculating the RMSD of the backbone atoms was used to check the stability of the four finalized ligand/receptor complexes. The amplitudes of the C atoms' fluctuation amplitudes were inversely proportional to the system's stability. The lower the RMSD variation, the more stable the system is, and the fluctuations of the C atoms in the system [41] will be smaller. The RMSD plot is used to understand the complex's stability, while the RMSF plot is used to understand its structural flexibility (Figure 6).

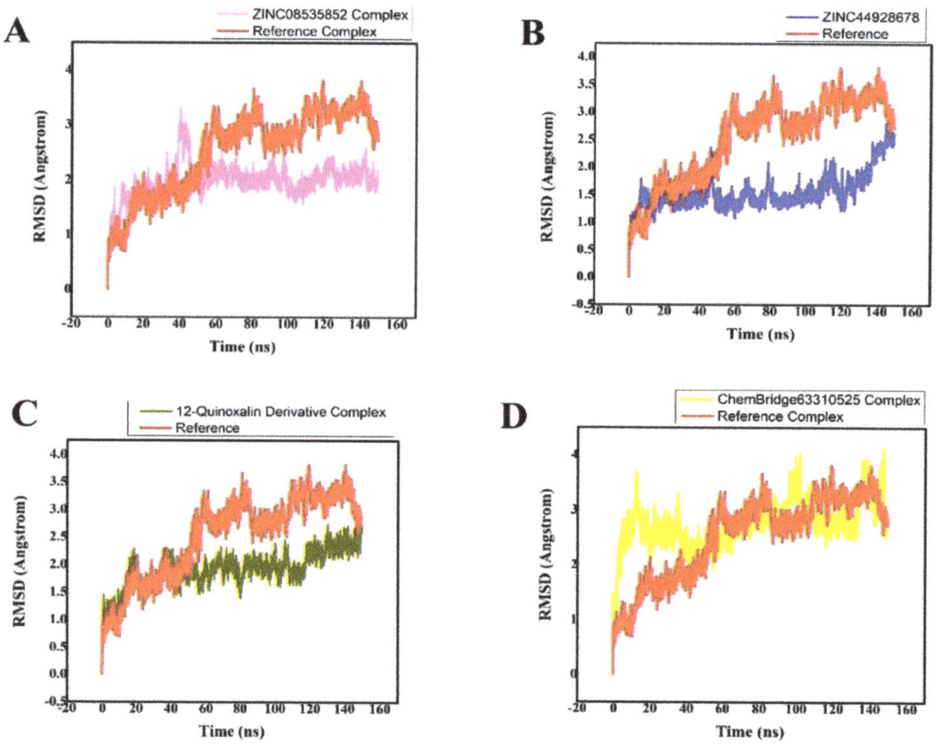

Figure 6. (**A–D**) Plot of the lead hits/M^{pro} and ref-ligand (N3)/M^{pro} complexes' root mean square deviations.

MDS equilibration was completed using the ZINC08535852/M^{pro} system at 150 ns. The RMSD graph showed that the amplitudes of the fluctuations rose with an RMSD score of 3 Å from 0 to 40 ns, and then the score decreased to 1.8 Å with slight fluctuations after 40 ns. After 140 ns up to 150 ns, the system was completely stabilized, with an RMSD of 1.8 Å and the least amount of fluctuations (Figure 6A).

The third compound is ZINC44928678 in complex with M^{pro}. This complex was then subjected to the MDS to produce an energy-minimized and stabilized structure. The system's RMSD graph revealed that the complex was stable from 0 to 100 ns of MDS at 1.5 Å, but that after MDS progression from 100 to 120 ns, their RMSD score increased to 2 Å as the system became unstable. The system was then stabilized after 120 ns, displaying smaller fluctuations with a slight increase in RMSD 2.5 Å, as shown in Figure 6B.

M^{pro} in complex with the 12-quinoxaline derivative was the second most active compound, with an RMSD of approximately 2.4 Å during the 150 ns simulation. The MDS of this ligand in contact with the M^{pro} can be shown in Figure 6C. This MDS was computed using RMSD scores of 2.4 Å, and the results are shown in Figure 6C. The RMSD graph of this system shows that the fluctuations became less and obtained an RMSD score of 2.4 Å after executing the 120–150 ns of MDS.

The ChemBridge63310525/M^{pro} (hit ligand/protein) complex's RMSD graph displayed the highest variations from 0 to 140 ns, with an increase of approximately 3.5 Å. Soon after 140 ns of MDS, smaller fluctuations were seen in the complex system's structural backbone, with an RMSD of approximately 2.8 Å (Figure 6D). M^{pro} undergoes conformational changes, as evidenced by the differences in the backbone RMSD in hits/M^{pro}.

The fluctuations of the protein residues upon ligand binding, were analyzed by root mean square fluctuation (RMSF) (Figure 7A–D). The stability of a complex can be inferred from its RMSF (root mean square fluctuation) trajectories. An unstable bond is indicated by a plot with a lot of movement. A low number, on the other hand, or less fluctuation, denotes well-structured and less distorted complicated regions. RMSF is an equilibrium property that is calculated from the entire MD trajectory, so is its time average. For M^{pro} with residues containing the four possible inhibitors, the RMSF of backbone atoms was computed. The active site residues that interacted with the ligands were discovered to be stable and showed minimal variations over time, confirming the stability of the molecules with the target protein. The main variations, in contrast, were associated with regions that were far from the ligand active site, and others were discovered in the flexible loop region. Additionally, it was noted that the RMSF variations in the complexes were smaller than those in the other complex, indicating that it had substantially less structural mobility than the other complex. All of the systems exhibited patterns that were strikingly similar, as presented in Figure 7A–D. Figure 7A–D clearly show the overall RMSF value for all complexes.

3.5. Radius of Gyration (Rg)

The relationship between Rg and time was plotted in order to determine the system's compactness. Low Rg values are explained by the structure's high stability and closed, compact structure, whereas high Rg values are explained by the structure's low compactness when compared to conformational entropy (more folded). Figure 8 shows the simulation Rg values for the four compounds, which are easily discernible (as shown). In the complex, the Rg values remained stable, indicating that the binding of these molecules does not alter the protein's structure. The retrieved four ligands/M^{Pro} complexes have shown low Rg values (compacted conformation of complexes) as compared to the reference ligand (N3)/M^{Pro}, therefore, indicating that the obtained hits were active inhibitors. The Rg is a parameter for assessing the compactness and overall dimension of the protein, which in turn signifies the folding and unfolding of the protein. The lower the gyration values, the more folded the protein is, and vice versa. Therefore, Rg was calculated to determine whether the compounds/ligands maintained the folding of the system. Condensed architecture and size are supported by the complexes' Rg range (Figure 8A–D).

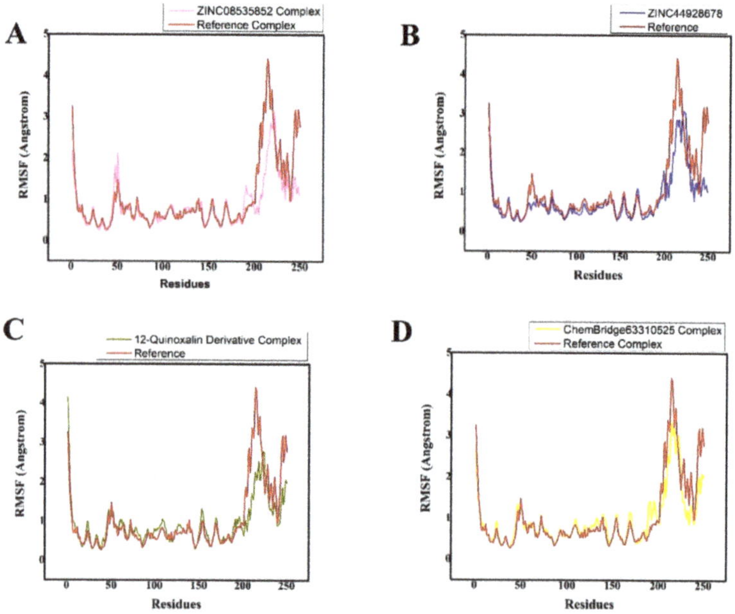

Figure 7. (**A–D**) Plot of the root mean square fluctuation of the identified lead hits/M^{pro} (purple, blue, olive green, and yellow, respectively) and ref-ligand (N3)/M^{pro} (red color) complexes. The various color schemes illustrated various hits/M^{pro} complexes and their related trajectories. Each system possessed the residual flexibility. On the graph, the number of residues is shown along the x-axis, and the RMSF is displayed along the y-axis in angstroms.

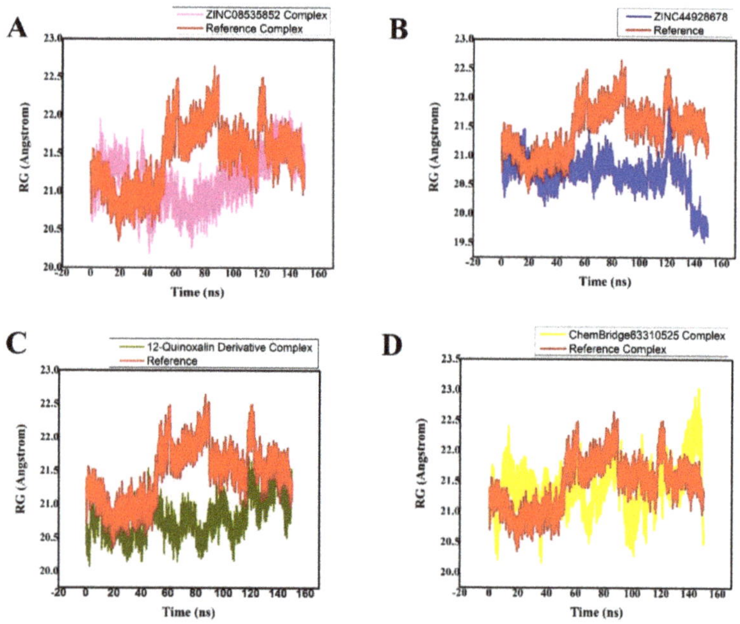

Figure 8. (**A–D**) The radii of gyration of the identified lead hits/M^{pro} (purple, blue, olive green, and yellow, respectively) and ref-ligand (N3)/M^{pro} (red color) complexes were plotted.

3.6. Molecular Mechanics with Generalized Born and Surface Area Solvation (MMGBSA)

This module also looked at the binding energy calculation of selected top compounds based on their affinity for the active site binding pocket. ZINC08535852, the first lead, had an MMGBSA score of −39.5546 Kcal/mol, whereas the active lead in in-house database, 12-quinoxaline derivative, had an even lower MMGBSA score of −37.8210 Kcal/mol. The MMGBSA binding free energies of the ZINC44928678 and ChemBridge63310525 are −35.8398 Kcal/mol and −33.2041 Kcal/mol, respectively. These results lead to the conclusion that the retrieved hits had a higher free binding energy score than the reference (−20.7812 Kcal/mol) (Table 3). In order to confirm the possible uses of these powerful inhibitors in COVID-19, we will soon design further experimental studies to test the inhibitory capacity of compounds towards SARS-CoV-2.

Table 3. The average backbone RMSD and binding free energies for the best four complexes. The data are reported as the average ± standard error of the mean (SEM).

Compound Names	RMSD (Å)	Binding Free Energies (Kcal/mol)
ZINC08535852 (ZINC database)	1.8 ± 0.005	−39.5546 ± 0.3671
ZINC44928678 (ZINC database)	2.5 ± 0.005	−35.8398 ± 0.1901
12-quinoxaline derivative in-house database	2.4 ± 0.005	−37.8210 ± 0.5091
ChemBridge63310525 (ChemBridge database)	2.8 ± 0.005	−33.2041 ± 0.2102
Reference (N3)	2.8 ± 0.0126	−20.7812 ± 0.4214

The interaction reports of the selected hits (ZINC08535852, ZINC44928678, 12-quinoxaline derivative, ChemBridge63310525, and N3 (reference ligand)) were mentioned after the MD simulations in Table 2, while the remaining data were after the docking, i.e., the docked complexes' interactions reports.

4. Conclusions

The primary goal of this study was to conduct a SBVS of compounds from the in-house, ZINC, and ChemBridge databases, as well as molecular docking and MDS of selected compounds and reference ligands (N3), as well as an estimation of binding interactions against the M^{pro} of SARS-CoV-2. Three active inhibitors, the top active one hit from each database, from the ZINC, ChemBridge, and in-house databases (ZINC08535852, ChemBridge63310525, and 12-quinoxaline derivative, respectively) showed a strong interaction against the active site of the M^{pro} of SARS-CoV-2. These compounds were identified as the most active anti-viral compounds ZINC08535852, ChemBridge63310525, and 12-quinoxaline derivative, against the M^{pro} of SARS-CoV-2. These results show that these compounds have the potential to be employed as a medication to treat SARS-CoV-2 illness. If in vitro and in vivo clinical testing are followed, the compounds could potentially serve as a candidate for the COVID-19 regimen.

Author Contributions: Conceptualization, M.G., M.U. and M.A.; Data curation, M.G., S.G., S.S.u.H., S.Q.A. and S.B.; Formal analysis, H.A.K. and M.A.; Investigation, M.S.; Methodology, H.A.K.; Project administration, S.G.; Writing—original draft, M.G., H.A.K., S.S.u.H., S.B., S.Q.A., M.S. and M.A.; Writing—review and editing, S.S.u.H., S.B., M.S. and S.G. All authors have read and agreed to the published version of the manuscript.

Funding: This research received no external funding.

Institutional Review Board Statement: Not applicable.

Informed Consent Statement: Not applicable.

Data Availability Statement: Data is contained within the article. The data presented in this study is available to researchers upon request.

Acknowledgments: The authors wish to thank the University of Oradea, Oradea, Romania, for its financial support in publishing this paper.

Conflicts of Interest: All authors declared that there are no conflicts of interest.

References

1. Team, E. The epidemiological characteristics of an outbreak of 2019 novel coronavirus diseases (COVID-19)—China, 2020. *China CDC Wkly.* **2020**, *2*, 113.
2. Gorbalenya, A.E.; Baker, S.C.; Baric, R.; Groot, R.J.d.; Drosten, C.; Gulyaeva, A.A.; Haagmans, B.L.; Lauber, C.; Leontovich, A.M.; Neuman, B.W. Severe acute respiratory syndrome-related coronavirus: The species and its viruses–a statement of the Coronavirus Study Group. *BioRxiv* **2020**, 1–15. [CrossRef]
3. Zhou, P.; Yang, X.-L.; Wang, X.-G.; Hu, B.; Zhang, L.; Zhang, W.; Si, H.-R.; Zhu, Y.; Li, B.; Huang, C.-L. A pneumonia outbreak associated with a new coronavirus of probable bat origin. *Nature* **2020**, *579*, 270–273. [CrossRef] [PubMed]
4. Wu, F.; Zhao, S.; Yu, B.; Chen, Y.-M.; Wang, W.; Song, Z.-G.; Hu, Y.; Tao, Z.-W.; Tian, J.-H.; Pei, Y.-Y. A new coronavirus associated with human respiratory disease in China. *Nature* **2020**, *579*, 265–269. [CrossRef] [PubMed]
5. Li, Y.; Tenchov, R.; Smoot, J.; Liu, C.; Watkins, S.; Zhou, Q. A comprehensive review of the global efforts on COVID-19 vaccine development. *ACS Cent. Sci.* **2021**, *7*, 512–533. [CrossRef] [PubMed]
6. Cui, J.; Li, F.; Shi, Z.-L. Origin and evolution of pathogenic coronaviruses. *Nat. Rev. Microbiol.* **2019**, *17*, 181–192. [CrossRef]
7. Jalal, K.; Khan, K.; Basharat, Z.; Abbas, M.N.; Uddin, R.; Ali, F.; Khan, S.A. Reverse vaccinology approach for multi-epitope centered vaccine design against delta variant of the SARS-CoV-2. *Environ. Sci. Pollut. Res.* **2022**, *29*, 60035–60053. [CrossRef]
8. Li, Q.; Wu, J.; Nie, J.; Zhang, L.; Hao, H.; Liu, S.; Zhao, C.; Zhang, Q.; Liu, H.; Nie, L. The impact of mutations in SARS-CoV-2 spike on viral infectivity and antigenicity. *Cell* **2020**, *182*, 1284–1294.e9. [CrossRef]
9. Plante, J.A.; Liu, Y.; Liu, J.; Xia, H.; Johnson, B.A.; Lokugamage, K.G.; Zhang, X.; Muruato, A.E.; Zou, J.; Fontes-Garfias, C.R. Spike mutation D614G alters SARS-CoV-2 fitness. *Nature* **2021**, *592*, 116–121. [CrossRef]
10. Ge, X.-Y.; Li, J.-L.; Yang, X.-L.; Chmura, A.A.; Zhu, G.; Epstein, J.H.; Mazet, J.K.; Hu, B.; Zhang, W.; Peng, C. Isolation and characterization of a bat SARS-like coronavirus that uses the ACE2 receptor. *Nature* **2013**, *503*, 535–538. [CrossRef]
11. Liu, C.; Zhou, Q.; Li, Y.; Garner, L.V.; Watkins, S.P.; Carter, L.J.; Smoot, J.; Gregg, A.C.; Daniels, A.D.; Jervey, S. Research and Development on Therapeutic Agents and Vaccines for COVID-19 and Related Human Coronavirus Diseases. *ACS Cent. Sci.* **2020**, *6*, 315–331. [CrossRef] [PubMed]
12. Zumla, A.; Chan, J.F.; Azhar, E.I.; Hui, D.S.; Yuen, K.-Y. Coronaviruses—Drug discovery and therapeutic options. *Nat. Rev. Drug Discov.* **2016**, *15*, 327–347. [CrossRef] [PubMed]
13. Shang, J.; Wan, Y.; Luo, C.; Ye, G.; Geng, Q.; Auerbach, A.; Li, F. Cell entry mechanisms of SARS-CoV-2. *Proc. Natl. Acad. Sci. USA* **2020**, *117*, 11727–11734. [CrossRef] [PubMed]
14. Ghosh, A.K.; Xi, K.; Grum-Tokars, V.; Xu, X.; Ratia, K.; Fu, W.; Houser, K.V.; Baker, S.C.; Johnson, M.E.; Mesecar, A.D. Structure-based design, synthesis, and biological evaluation of peptidomimetic SARS-CoV 3CLpro inhibitors. *Bioorg. Med. Chem. Lett.* **2007**, *17*, 5876–5880. [CrossRef]
15. Wu, C.; Liu, Y.; Yang, Y.; Zhang, P.; Zhong, W.; Wang, Y.; Wang, Q.; Xu, Y.; Li, M.; Li, X. Analysis of therapeutic targets for SARS-CoV-2 and discovery of potential drugs by computational methods. *Acta Pharm. Sin. B* **2020**, *10*, 766–788. [CrossRef]
16. Zhavoronkov, A. Medicinal chemists versus machines challenge: What will it take to adopt and advance artificial intelligence for drug discovery? *J. Chem. Inf. Model.* **2020**, *60*, 2657–2659. [CrossRef] [PubMed]
17. Narayanan, A.; Narwal, M.; Majowicz, S.A.; Varricchio, C.; Toner, S.A.; Ballatore, C.; Brancale, A.; Murakami, K.S.; Jose, J. Identification of SARS-CoV-2 inhibitors targeting Mpro and PLpro using in-cell-protease assay. *Commun. Biol.* **2022**, *5*, 169. [CrossRef]
18. Shams ul Hassan, S.; Abbas, S.Q.; Hassan, M.; Jin, H.-Z. Computational Exploration of Anti-Cancer Potential of GUAIANE Dimers from Xylopia vielana by Targeting B-Raf Kinase Using Chemo-Informatics, Molecular Docking, and MD Simulation Studies. *Anti-Cancer Agents Med. Chem.* **2022**, *22*, 731–746. [CrossRef]
19. Biswas, D.; Nandy, S.; Mukherjee, A.; Pandey, D.; Dey, A. Moringa oleifera Lam. and derived phytochemicals as promising antiviral agents: A review. *S. Afr. J. Bot.* **2020**, *129*, 272–282. [CrossRef]
20. Keretsu, S.; Bhujbal, S.P.; Cho, S.J. Molecular modeling studies of pyrrolo [2,3-d] pyrimidin-4-amine derivatives as JAK1 inhibitors based on 3D-QSAR, molecular docking, molecular dynamics (MD) and MM-PBSA calculations. *J. Biomol. Struct. Dyn.* **2021**, *39*, 753–765. [CrossRef]
21. Zhou, T.; Wu, Z.; Das, S.; Eslami, H.; Müller-Plathe, F. How Ethanolic Disinfectants Disintegrate Coronavirus Model Membranes: A Dissipative Particle Dynamics Simulation Study. *J. Chem. Theory Comput.* **2022**, *18*, 2597–2615. [CrossRef] [PubMed]
22. Jakalian, A.; Jack, D.B.; Bayly, C.I. Fast, efficient generation of high-quality atomic charges. AM1-BCC model: II. Parameterization and validation. *J. Comput. Chem.* **2002**, *23*, 1623–1641. [CrossRef]
23. Ghufran, M.; Khan, H.A.; Ullah, M.; Ghufran, S.; Ayaz, M.; Siddiq, M.; Hassan, S.S.u.; Bungau, S. In Silico Strategies for Designing of Peptide Inhibitors of Oncogenic K-Ras G12V Mutant: Inhibiting Cancer Growth and Proliferation. *Cancers* **2022**, *14*, 4884. [CrossRef]

24. Ghufran, M.; Rehman, A.U.; Ayaz, M.; Ul-Haq, Z.; Uddin, R.; Azam, S.S.; Wadood, A. New lead compounds identification against KRas mediated cancers through pharmacophore-based virtual screening and in vitro assays. *J. Biomol. Struct. Dyn.* **2022**, 1–15. [CrossRef] [PubMed]
25. Hassan, S.S.; Tiwari, S.; Guimarães, L.C.; Jamal, S.B.; Folador, E.; Sharma, N.B.; de Castro Soares, S.; Almeida, S.; Ali, A.; Islam, A. Proteome scale comparative modeling for conserved drug and vaccine targets identification in Corynebacterium pseudotuberculosis. *BMC Genom.* **2014**, *15*, S3. [CrossRef]
26. Ghufran, M.; Rehman, A.U.; Shah, M.; Ayaz, M.; Ng, H.L.; Wadood, A. In-silico design of peptide inhibitors of K-Ras target in cancer disease. *J. Biomol. Struct. Dyn.* **2020**, *38*, 5488–5499. [CrossRef]
27. Taj, S.; Ahmad, M.; Alshammari, A.; Alghamdi, A.; Ashfaq, U.A. Exploring the therapeutic potential of benzothiazine-pyrazole hybrid molecules against alpha-glucosidase: Pharmacological and molecular modelling based approach. *Saudi J. Biol. Sci.* **2022**, *29*, 1416–1421. [CrossRef]
28. Maier, J.A.; Martinez, C.; Kasavajhala, K.; Wickstrom, L.; Hauser, K.E.; Simmerling, C. ff14SB: Improving the accuracy of protein side chain and backbone parameters from ff99SB. *J. Chem. Theory Comput.* **2015**, *11*, 3696–3713. [CrossRef] [PubMed]
29. Zwanzig, R. Nonlinear generalized Langevin equations. *J. Stat. Phys.* **1973**, *9*, 215–220. [CrossRef]
30. Ryckaert, J.-P.; Ciccotti, G.; Berendsen, H.J. Numerical integration of the cartesian equations of motion of a system with constraints: Molecular dynamics of n-alkanes. *J. Comput. Phys.* **1977**, *23*, 327–341. [CrossRef]
31. Gotz, A.W.; Williamson, M.J.; Xu, D.; Poole, D.; Le Grand, S.; Walker, R.C. Routine microsecond molecular dynamics simulations with AMBER on GPUs. 1. Generalized born. *J. Chem. Theory Comput.* **2012**, *8*, 1542–1555. [CrossRef]
32. Sun, H.; Li, Y.; Tian, S.; Xu, L.; Hou, T. Assessing the performance of MM/PBSA and MM/GBSA methods. 4. Accuracies of MM/PBSA and MM/GBSA methodologies evaluated by various simulation protocols using PDBbind data set. *Phys. Chem. Chem. Phys.* **2014**, *16*, 16719–16729. [CrossRef] [PubMed]
33. Miller, B.R., III; McGee, T.D., Jr.; Swails, J.M.; Homeyer, N.; Gohlke, H.; Roitberg, A.E. MMPBSA. py: An efficient program for end-state free energy calculations. *J. Chem. Theory Comput.* **2012**, *8*, 3314–3321. [CrossRef]
34. Hui, D.S.; Azhar, E.I.; Madani, T.A.; Ntoumi, F.; Kock, R.; Dar, O.; Ippolito, G.; Mchugh, T.D.; Memish, Z.A.; Drosten, C. The continuing 2019-nCoV epidemic threat of novel coronaviruses to global health—The latest 2019 novel coronavirus outbreak in Wuhan, China. *Int. J. Infect. Dis.* **2020**, *91*, 264–266. [CrossRef]
35. Wang, F.; Chen, C.; Tan, W.; Yang, K.; Yang, H. Structure of main protease from human coronavirus NL63: Insights for wide spectrum anti-coronavirus drug design. *Sci. Rep.* **2016**, *6*, 22677. [CrossRef] [PubMed]
36. Maia, E.H.B.; Assis, L.C.; De Oliveira, T.A.; Da Silva, A.M.; Taranto, A.G. Structure-based virtual screening: From classical to artificial intelligence. *Front. Chem.* **2020**, *8*, 343. [CrossRef]
37. Lipinski, C.A.; Lombardo, F.; Dominy, B.W.; Feeney, P.J. Experimental and computational approaches to estimate solubility and permeability in drug discovery and development settings. *Adv. Drug Deliv. Rev.* **1997**, *23*, 3–25. [CrossRef]
38. Lipinski, C.A. Drug-like properties and the causes of poor solubility and poor permeability. *J. Pharmacol. Toxicol. Methods* **2000**, *44*, 235–249. [CrossRef] [PubMed]
39. Mahgoub, R.E.; Mohamed, F.E.; Alzyoud, L.; Ali, B.R.; Ferreira, J.; Rabeh, W.M.; AlNeyadi, S.S.; Atatreh, N.; Ghattas, M.A. The Discovery of Small Allosteric and Active Site Inhibitors of the SARS-CoV-2 Main Protease via Structure-Based Virtual Screening and Biological Evaluation. *Molecules* **2022**, *27*, 6710. [CrossRef]
40. Kumar, P.; Choonara, Y.E.; Pillay, V. In silico affinity profiling of neuroactive polyphenols for post-traumatic calpain inactivation: A molecular docking and atomistic simulation sensitivity analysis. *Molecules* **2014**, *20*, 135–168. [CrossRef]
41. Li, W.-Y.; Duan, Y.-Q.; Ma, Y.-C.; Lu, X.-H.; Ma, Y.; Wang, R.-L. Exploring the cause of the inhibitor 4AX attaching to binding site disrupting protein tyrosine phosphatase 4A1 trimerization by molecular dynamic simulation. *J. Biomol. Struct. Dyn.* **2019**, *37*, 4840–4851.

bioengineering

MDPI

Article

Modelling and Analysis of Hybrid Transformation for Lossless Big Medical Image Compression

Xingsi Xue, Raja Marappan, Sekar Kidambi Raju, Rangarajan Raghavan, Rengasri Rajan and Osamah Ibrahim Khalaf et al.

[1] Fujian Provincial Key Laboratory of Big Data Mining and Applications, Fujian University of Technology, Fuzhou 350011, China
[2] School of Computing, SASTRA Deemed University, Thanjavur 613401, India
[3] Department of Solar, Al-Nahrain Renewable Energy Research Center, Al-Nahrain University, Baghdad 64040, Iraq
[4] Department of Computer Engineering, University of Technology, Baghdad 19006, Iraq
* Correspondence: sekar_kr@cse.sastra.ac.in; Tel.: +91-9843710643

Abstract: Due to rapidly developing technology and new research innovations, privacy and data preservation are paramount, especially in the healthcare industry. At the same time, the storage of large volumes of data in medical records should be minimized. Recently, several types of research on lossless medically significant data compression and various steganography methods have been conducted. This research develops a hybrid approach with advanced steganography, wavelet transform (WT), and lossless compression to ensure privacy and storage. This research focuses on preserving patient data through enhanced security and optimized storage of large data images that allow a pharmacologist to store twice as much information in the same storage space in an extensive data repository. Safe storage, fast image service, and minimum computing power are the main objectives of this research. This work uses a fast and smooth knight tour (KT) algorithm to embed patient data into medical images and a discrete WT (DWT) to protect shield images. In addition, lossless packet compression is used to minimize memory footprints and maximize memory efficiency. JPEG formats' compression ratio percentages are slightly higher than those of PNG formats. When image size increases, that is, for high-resolution images, the compression ratio lies between 7% and 7.5%, and the compression percentage lies between 30% and 37%. The proposed model increases the expected compression ratio and percentage compared to other models. The average compression ratio lies between 7.8% and 8.6%, and the expected compression ratio lies between 35% and 60%. Compared to state-of-the-art methods, this research results in greater data security without compromising image quality. Reducing images makes them easier to process and allows many images to be saved in archives.

Keywords: big data; data security; knight tour; steganography; wavelet transform; lossless compression

Citation: Xingsi Xue, Raja Marappan, Sekar Kidambi Raju, Rangarajan Raghavan, Rengasri Rajan and Osamah Ibrahim Khalaf et al. Modelling and Analysis of Hybrid Transformation for Lossless Big Medical Image Compression. *Bioengineering* **2023**, *10*, 333. https://doi.org/10.3390/bioengineering10030333

Academic Editor: Yunfeng Wu

Received: 28 January 2023
Revised: 15 February 2023
Accepted: 27 February 2023
Published: 6 March 2023

1. Introduction

In recent years, there has been a need for extensive medical data information security in all research fields. One such field where information security is crucial in medicine and pharmacology. Today, the number of diseases and people affected by them has increased, posing a potential challenge in storing various medically significant data in limited storage spaces. In addition, data security cannot be compromised. Over the years, steganography techniques have proven to be one set of effective methods to maintain the privacy of big medical data. This method uses an overlay image to embed and retrieve data algorithmically. There are several options for entering information into big medical data. One such variation used in this paper is the knight tour (KT) traversal technique combined with least significant bit (LSB) substitution to preserve data. The big data of the patient are added to the medical image, which is the cover image of this steganography technique. The KT performs the embedding, and the binary information is embedded in the LSB position of the medical big data image pixels placed in the KT path. To increase security, the discrete wavelet transform (DWT) is applied to the cover image to make it unreadable, and lossless wavelet compression is used to efficiently store the transformed image in an extensive medical data repository. This hybrid approach ensures the security of patient information and the efficient storage of information in a medical database using as little storage space as possible.

Initial attempts to implement communications, photo archiving systems and teleradiology applications have encountered significant obstacles. Large picture files present problems for both transmission and storage due to the high cost of the equipment required to manage them. Lossless compression techniques only accomplish 2:1 to 3:1 reductions for medical photos by using redundancy within an image to transfer image information more effectively while permitting flawless reconstruction. Images can be reduced by arbitrarily huge ratios using irreversible or "lossy" processes, but the original pictures are not fully replicated. However, the reproduction quality could be sufficient to prevent visible picture deterioration and diminished diagnostic usefulness. This paper examines wavelet compression in a new light. We compare the method to the famous Joint Photographic Experts Group (JPEG) approach. We review methods for measuring compression algorithms' effectiveness and look at new developments in wavelet compression.

The motivations behind this research work, along with the significant contributions respecting processes, are as follows:

- Compressing the image—Apply wavelet analysis to separate and approximate the image information based on the sub-images and sub-signals.
- Pixel approximation for compression—Apply the images' horizontal, vertical, and diagonal features and approximate the sub-signals for better pixel approximation.
- Frequency and time localization—Apply WT effectively to reduce computational complexity.
- Signal-features-based compression—Apply WT to distinguish between the minute signal features during the compression.
- Quantization and lossless compression—Apply the hybrid transformation to quantize the coefficients to optimize the lossless image compression during the transmission.

Wavelet compression and JPEG fall under the "transform-based lossy compression algorithms" umbrella category. Transformation, quantization, and encoding are the three processes that make up these approaches. A picture is converted from grayscale values in the spatial domain to coefficients in another domain in a lossless procedure known as transformation. The Fourier transform, used to recreate magnetic resonance (MR) pictures, is one well-known transform. The DCT and WT are two more transformations more often employed in compression. The transformation phase does not result in any information loss. Information is lost during the quantization process. While less significant coefficients are loosely approximated, frequently as zero, efforts are made to preserve the more significant coefficients. Converting floating points to decimals might be enough to perform quantization: values are converted to integers. The quantized coefficients are finally encoded. The quantized coefficients are compactly represented in this step, which is also lossless, to allow for adequate image storage or transmission.

The paper is organized as follows. Section 2 presents a literature survey, Section 3 discusses the proposed model, and Section 4 focuses on the results and analysis. The conclusions and a consideration of future enhancements are drawn in Section 5.

2. Literature Survey and Critiques

Various steganography methods and standard guidelines have recently been proposed that provide valuable recommendations and suggest using object-oriented mechanisms [1]. A three-layer technique, including pixel statistics preservation, Moore space-filling curves, and Hilbert space-filling curves, increases and improves security [2]. A detailed comparative analysis was made of different digital steganography techniques, and performance was evaluated in terms of peak signal-to-noise ratio (PSNR) and mean square error (MSE) values [3]. Image compression of black and white images using a wavelet transform that uses the LGB algorithm and an error-correction method to minimize distortion has been presented. The compression result was compared with other techniques for encoding DWT-coupled map grid (CML) data [4]. The results show that the proposed encryption algorithm has the advantage of large key space, high security, and fast encryption or decryption speed [5]. A recently developed generic codec supporting JPEG 2000 with volumetric extension (JP3D) was used to investigate how to optimally compress volumetric medical images using JP3D. Various directional WTs and a general in-group prediction mode provide guidelines and settings for optimal compression of medical volumetric images at an acceptable level of complexity [6].

A low-complexity 2D image compression method was tested using Haar waves, and compressed image quality was obtained. Some factors, such as compression ratio (CR), PSNR, mean opinion score (MOS), picture quality scale (PQS), etc., have been used to evaluate the quality of compressed images [7]. The work used a simple LSB substitution technique for data hiding. The MSE is calculated, and the resulting image has no significant changes from the original image [8]. An advanced steganography technique that includes side information to calculate the correlations between neighboring pixels and estimate degrees of smoothness has been used. The results showed that the technology provides a high delivery capacity with less distortion [9]. Steganography was used by adjusting the LSB value of the color pixel intensity by simple binary addition. The results showed that the embedding power was twice that of the traditional technique [10]. The methods used were DWT for medical image compression. Research shows that correlation and redundancy are reduced in the DWT domain, while random permutation of pixels with an encryption key leads to confusion and dispersion [11]. Information hiding was developed with multi-pixel difference (MPD) and LSB (least significant bit) replacement to improve image quality and increase information storage capacity. The results showed that the improved image and resolution were undistorted [12]. Concatenative singular value decomposition (SVD) and optimized DWT were used. Huffman encoding and decoding were also performed.

The results were compared with other image compression methods based on CR, PSNR, SSIM, and MSE [13]. The various scenography and digital watermarking techniques for beginners were explored and acted as a guide to help understand the concepts and apply them very easily [14]. The coefficients of the applied 2-D orthogonal WT and the transformed image were coded and quantified according to the local estimated noise sensitivity. The human visual system offers a high degree of compression [15]. In image compression, fusion, and encryption, big medical data CS, chaos, and fractional Fourier transforms were used simultaneously. The technique was developed to reduce data and simplify the keys [16]. Secure and lossless digital image watermarking was developed based on DWT and DCT databases to preserve patient privacy in a medical database. Performance development factors include PSNR measurements and their correlations with overall image degradation [17]. DWT was used for the fast and secure transmission of primary images online. The results showed that the algorithms quickly and securely send essential images in a group [18]. A wavelet-based approach was developed to compress and encrypt fused images by selecting salient and less salient information. Fusion is performed by error measurement with compression, and encryption methods use pseudo-random number sequences and Huffman coding. The results showed that the proposed method was better than all others [19]. DWT was used for medical big data image compression. The results showed that the proposed algorithm has better security and performance than previous works [20]. This research used DCT, DWT, and watermark methods to store patient data. The results showed that the proposed system performs better un-detectability [21]. The modulation function was applied, and many advantages were achieved regarding good resolution, overcoming LSB compensation, and creating large storage space to achieve a sizeable hidden volume [22]. The Embedded Zero tree Wavelet (EZW) algorithm provides an efficient technique for low bit rates and high compression and advanced EZW techniques for lossless and lossless grayscale and color image compression [23]. The simple and optimized LSB replacement method and genetic algorithm were developed for image embedding. The results showed that the embedded image was not significantly affected, and the hiding strategy was improved [24]. The new type of steganography using the LSB replacement and pixel value difference (PVD) methods was developed to improve image quality and maximize the hidden space [25–28].

The entropy, run length, and dictionary-based compression techniques were developed to achieve lossless compression [29–33]. The DNN and principal component analysis strategies were applied to predict data compression using entropy values. The images were divided into blocks, and then applied compression [34].

3. Proposed Methodology

The proposed method mainly focuses on KT steganography, WTs, and compression. Today, image and audio file sizes are drastically compressed in a new mathematical way. Lossless compression is an old technology, but JPEG and MP3 now use lossless compression and are essentially required in the new environment. The original document was extensively searched using a mathematical model. Typically, compression up to a tenth of the original image length is possible. The quality and accuracy never deteriorate. Record compression and image compression must not allow images to be lost. The length of the original image must match the load image after it is translated.

The fast KT approach is designed to move only in the L direction, i.e., three down and one to the right, starting at the first pixel of the image. The cover image is converted using DWT, and the "wpdencmp" lossless wavelet packet based on the "Haar" wavelet is used to compress the transformed image. The complete methodology is described in the architecture diagram (Figure 1). DWT has wide applications in technological understanding, engineering, mathematics, and computer engineering knowledge. In particular, it encodes characters to symbolize separate records in an extra-redundant format, often as a prerequisite for fast compression. Both strategies have their advantages and disadvantages. Like DWT, it offers a higher compression ratio of 1:3 without losing photos but still requires more processing power. DCT can perform electrical processing, and artifact blocks are like losing some statistics' continuous WTs.

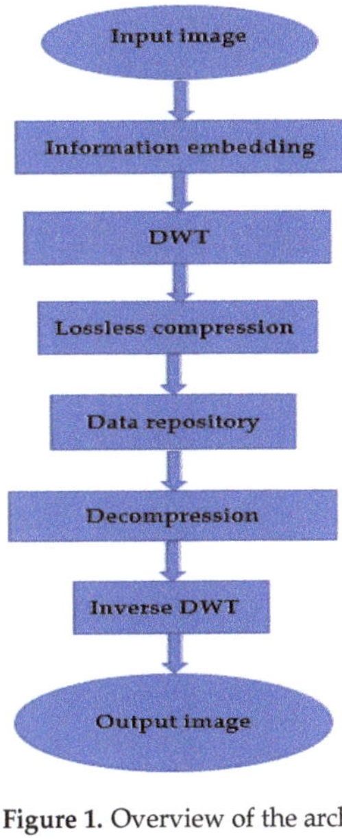

Figure 1. Overview of the architecture.

3.1. Architecture Diagram

Image compression is a method used to reduce image space. In decompression, a small part of the image is first removed and scaled to the large image. The primary purpose of image compression is to reduce minor images and noise because images are removed. Image redundancy is squared to make the image effective. This technique reduces the bit size of an image without affecting image quality. Reducing the image quality does not affect the image quality of any account. The shorter the length, the more images can be stored under the archive. The disk space required to store images is small or large.

3.2. Proposed Algorithm

The clinical data repository combines statistics from various clinical resources, including EMR or laboratory systems, to provide a complete picture of a patient's care. These archives are characterized as databases containing scientific statistics. Healthcare extensive records is a term used to describe the large number of records created by adopting digital technologies that collect patient statistics and help manage healthcare services that are otherwise too large and complex for classic technologies. Clinical repositories can provide a rich overview of patients, their clinical conditions, and outcomes. The database can provide a way to explore associations and capacity styles between disease progression and treatment. The proposed algorithm with improved embedding is sketched in Algorithm 1.

The DWT coefficients were first tested (approximate and detail coefficients are processed separately) by inserting zeros between each coefficient, doubling each length exactly. Just as the concept of filter financial institutions can determine DWT, so can IDWT reconstruction be performed by taking the real first N/2-1 coefficients of the DWT coefficients and adding them to the transfer. RMSE suggests that square deviation is one of the most typically used measures for comparing excellent predictions. It shows how some distance predictions fall from measured actual values using Euclidean distance. To compute RMSE, calculate the residual (the distinction between prediction and fact) for each statistical factor, compute the residual norm for every information factor, compute the suggestion of residuals, and take the square root of that mean. RMSE is typically utilized in supervised mastering applications, as RMSE uses and requires accurate measurements for each predicted record factor.

Algorithm 1: Improved Embedding

1: Load the medical image and the patient data as X ← medical image and D ← patient data.
2: Adjust the matrix dimensions such that the rows are multiples of 3 and the columns of 2.
3: Pad the extra row or column with the value '256' as the pixel ranges between 0 and 255.
4: Extract the ASCII value of D and convert it to its equivalent binary value as D → ASCII (D) → Binary(D).
5: Apply KT for pixel traversing the image in the 'L' path, and change the LSB bit of each pixel with each bit of the binary value from D.
6: Delete the extra padded row or column.
7: Extract the red, green, and blue components from X.
8: Apply forward DWT for the red, green, and blue components using the formula

$$I_{DWT}(a,b) = 2^{-a/2}\int_{-\infty}^{+\infty} i(t)\psi(2^{-a}t-b)dt \quad (1)$$

$$\psi_{a,b}(t) = 2^{-a/2}\psi(2^{-a}t-b) \quad (2)$$

where ψ(t) is the mother wavelet.
9: Combine the individual transformed components into one matrix.
trans = combine (red, blue, and green)
10: Compute lossless wavelet packet compression using 'wpdencmp' function with 'haar' wavelet packet and store it in the repository.
Z = compress (wpdencmp, trans, haar)

The extraction algorithm is defined in Algorithm 2 as follows.

Algorithm 2: Extraction

1: Retrieve the compressed image from the medical database and perform wavelet decompression.
trans = decompress (Z)
2: Apply Inverse DWT (IDWT).

$$i(t) = \sum_a \sum_b I_{DWT}(a,b)\left[2^{-a/2}\psi(2^{-a}t - b)\right] \tag{3}$$

3: Extract the red, blue, and green components from the reconstructed image, and combine them.
4: Make the image matrix traversable by adding an extra row or column and pad with the value '256'.
5: Apply the KT to extract the LSB values from the pixels of the image traversing in an 'L' pattern.
6: Delete the extra padded row or column.
7: Convert the obtained binary bits to ASCII values and the character.
Binary (D) = ASCII (D) = char (D) = D
8: Update the patient data = D and medical image = X.

The proposed image compression is defined in Algorithm 3.

3.3. Computational Performance of Wavelet Transform

The size of the filter set and the number of pixels in the image impact how quickly wavelet compression works. Although JPEG speed is related to the number of pixels, the DWT has substantially higher memory needs since the entire image must be changed simultaneously. In light of this, wavelet compression speed on a particular platform will be equivalent to JPEG compression speed for small pictures that can be stored entirely in memory but slower for bigger images. As an electronic file, the number of radiologic tests varies little between treatment techniques. A computed tomographic (CT) picture is 189 MB in size. A study with 40 images takes up roughly 16 MB for a two-view radiographic study and 20 MB for a study with 40 images. A typical MR picture is 8 MB; however, because many images are often collected, MR imaging examinations typically consume 10–15 MB. A cross-sectional picture may be decompressed using Windows NT 4.0 software (Microsoft, Redmond, Wash) in nanoseconds on a 200 MHz Pentium Pro computer with 64 MB of random access memory. A 10 MB radiograph may be decompressed using wavelet compression in around 9 s, compared to 7 s using the JPEG method. Standard modems may often operate via telephone lines at 28.8 kbit/sec rates. A 15 MB file would take roughly 90 min to transfer. A 10:1 compression ratio would cut the transmission time to 9 min, and a 33:1 ratio would lower it to 3 min. The improvement in transmission time is directly proportionate to the compression ratio employed. The compression ratio also linearly reduces the storage needed; for example, a 15:1 compression ratio would reduce the 15 MB research to just 1 MB [25–32].

Algorithm 3: Proposed Image Compression

1: Apply the improved embedding to the medical image.
2: Perform the extraction operation.
3: Apply the hard threshold using

$$y_{hard}(t) = x(t) \ when \ abs(x(t)) > T$$

$$y_{hard}(t) = 0, \quad when \ abs(x(t)) \leq T$$

4: Perform the soft threshold using

$$y_{soft}(t) = x(t) - T \ when \ abs(x(t)) > T$$

$$y_{soft}(t) = 0, \quad when \ abs(x(t)) \leq T$$

5: Find the entropy encoding h(s).

$$h(s) = -\sum_{j=1}^{q} p(s_i) log \ p(s_i)$$

6: Determine the Shannon entropy M(C(b))).
7: Apply the wavelets based multi resolution analysis.
8: Define the Haar transform
$I_{k,\ n} = [k2^{-n}, (k+1)2^{-n}]$
9: Perform the dilation using

$$\varnothing(x) = \sqrt{2}\sum_{k \in z} \varnothing(-k+2x)h(k)$$

10: Apply the wavelet operation using

$$\varphi(x) = \sqrt{2}\sum_{k \in z} g(-k+2x)g(k)$$

11: Perform filtering operations using orthogonal quadrature.

$$h(i) = \sum_{j=-\infty}^{\infty} u(j).h(-j+2i)$$

$$g(i) = \sum_{j=-\infty}^{\infty} u(j).g(-j+2i)$$

12: Apply the filters using bi-orthogonal quadrature.
13: Decompose the image using wavelets and apply the color conversion.

$$\begin{bmatrix} Y \\ C_b \\ C_r \end{bmatrix} = \begin{bmatrix} 0.3 & 0.6 & 0.1 \\ -0.2 & -0.3 & 0.5 \\ 0.5 & -0.4 & -0.1 \end{bmatrix} \begin{bmatrix} R \\ G \\ B \end{bmatrix}$$

14: Find RGB using

$$\begin{bmatrix} R \\ G \\ B \end{bmatrix} = \begin{bmatrix} 1 & 0.1 & 1.3 \\ 1 & 0.3 & -0.5 \\ 1 & 1.8 & -0.1 \end{bmatrix} \begin{bmatrix} Y \\ C_b \\ C_r \end{bmatrix}$$

15: Apply the lossless compression through

$$\begin{bmatrix} Y_r \\ V_r \\ U_r \end{bmatrix} = \begin{bmatrix} (2G+R+B)/4 \\ R-G \\ B-G \end{bmatrix}$$

$$\begin{bmatrix} G \\ R \\ B_r \end{bmatrix} = \begin{bmatrix} Y_r - (V_r+U_r)/4 \\ G+V_r \\ G+U_r \end{bmatrix}$$

4. Simulation and Analysis

The overall methodology of this research includes KT steganography, DWTs, and lossless wavelet packet compression of medical images for privacy, preservation, and efficient storage. The steganography technique uses the LSB replacement technique, which replaces LSB bits of pixels with patient data in binary form.

4.1. Datasets

The proposed model was simulated using datasets from http://www.aylward.org/notes/open-access-medical-image-repositories, accessed on 1 January 2023 [20–24]. The proposed model was tested using 43,752 chest radiographs from the National Institutes of Health Clinical Center (NIHCC) database. This comprises 1,00,000 images with relevant details, diagnostic information, and other publicly available datasets. The model was tested for various image sizes. The expected results were obtained with various data fixes and changes, which challenged the embedding algorithm, and additional data were included to verify the effectiveness of the compression algorithm.

4.2. Results and Analysis

The results of the steganography technique are shown in Figure 2; the changes in both the cover image and the steganography image cannot be distinguished. This demonstrates the superiority of the LSB replacement method in the field of steganography. The resulting steganography image is then transformed to render the image unreadable, and compression techniques are used to store the medical image in a database properly. This scenario is illustrated in Figure 3. This diagram shows the order of conversion, compression, decompression, and inverse conversion processes. To realize this methodology, we took as input cerebral hemorrhage steganography images already embedded in patient data and applied the DWT to obtain the transformed images.

The "dwt2" method is used for the conversion. In the DWT process, the original image is decomposed into up to two levels using the "Haar" wavelet. This decomposition produces horizontal, vertical, diagonal, and proximity components. The decomposed components are reconstructed using the DIWT to recover the original image. This strategy is illustrated in Figure 2, and the compression method takes a decomposed image as input and compresses it using a wavelet packet compression technique using "Haar" wavelet packets. The compression method, "wpdencmp," uses a soft thresholding technique that uses wavelet packets to compress the image and compute the threshold. This compression concept is illustrated in Figure 3, along with histograms of the original and compressed images. A step-by-step compaction process is shown in Figure 4. At each level, the images are refined, and the differences between levels are visible. The higher the number of coding levels of compression, the higher the image's compression ratio and recovered energy. The compressed image is decompressed by wavelet packet reconstruction using the accounting matrix values of the decomposed image. Horizontal, vertical, diagonal, and proximity components are extracted from the decompressed image, and an IDWT is applied to recover the original image. A title image embedded in the patient data is visualized.

In Figure 3, the steganography image of the brain hemorrhage is taken, and DWT is applied. After compression, the compressed image is displayed as a bar graph. The wavelet reconstruction method recovers the compressed image from the compressed image. The target image is retrieved from the decompressed image by applying the IDWT method to the previous step. Figure 4 represents the DWT and IDWT of the brain hemorrhage image. A sample segmentation of the transformed image is also displayed.

Figure 5 represents the histograms of the steganography and decomposed and compressed images in red, blue, and pink colors, respectively. The obtained outcome shows that the adopted methodology is efficient regarding high-resolution lossless compression of medical images. In the resultant images, the novelty of the work is apparent; it can be seen that the proposed methodology enables excellent elaboration and enhancement using improved WTs and lossless compressions. Figure 6 represents the transition steps of the wavelet packet compression technique, where the images are clearly distinguished from one another.

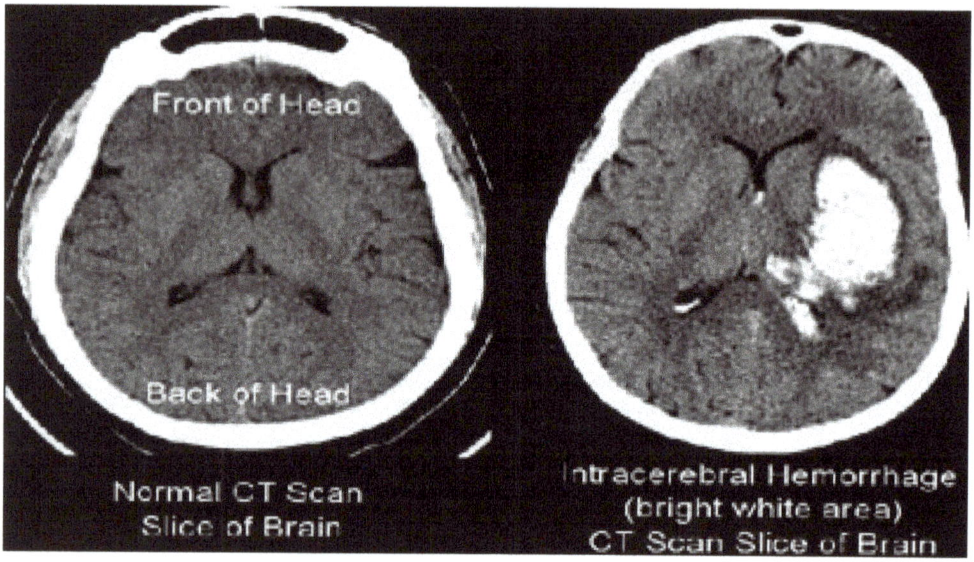

Figure 2. The embedded image.

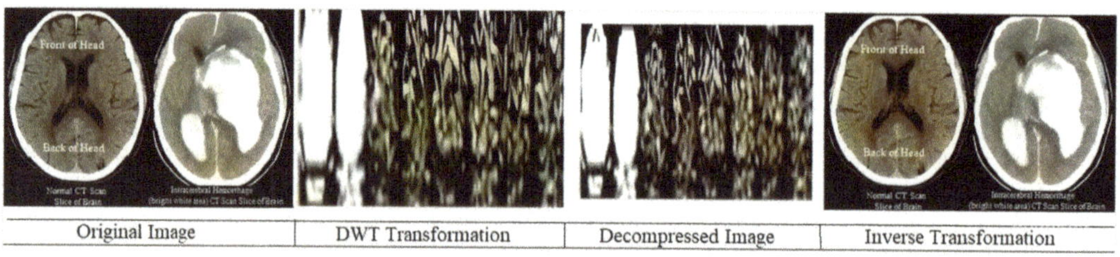

Figure 3. Hybrid transformation and compression.

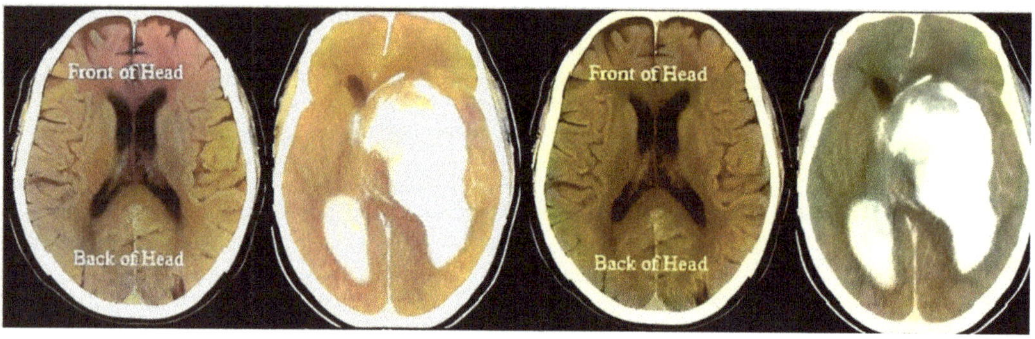

Figure 4. Applying DWT.

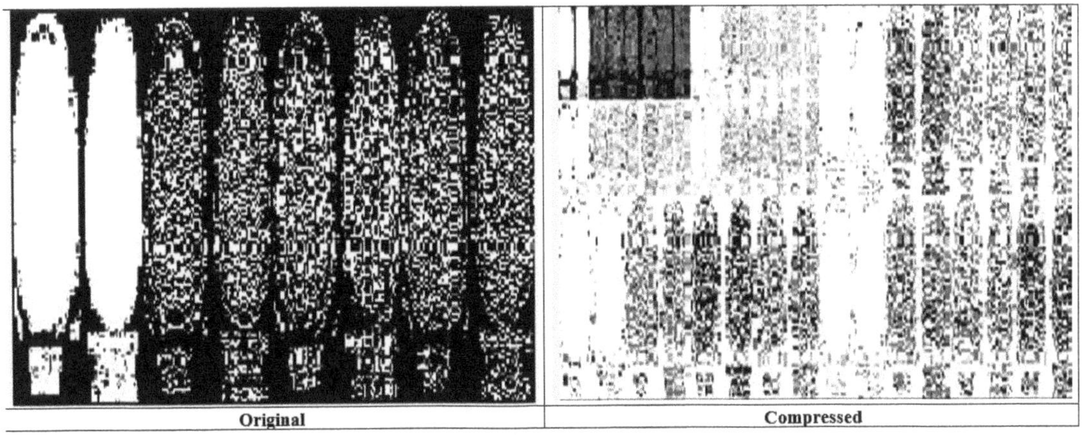

Figure 5. Histograms of original and compressed images.

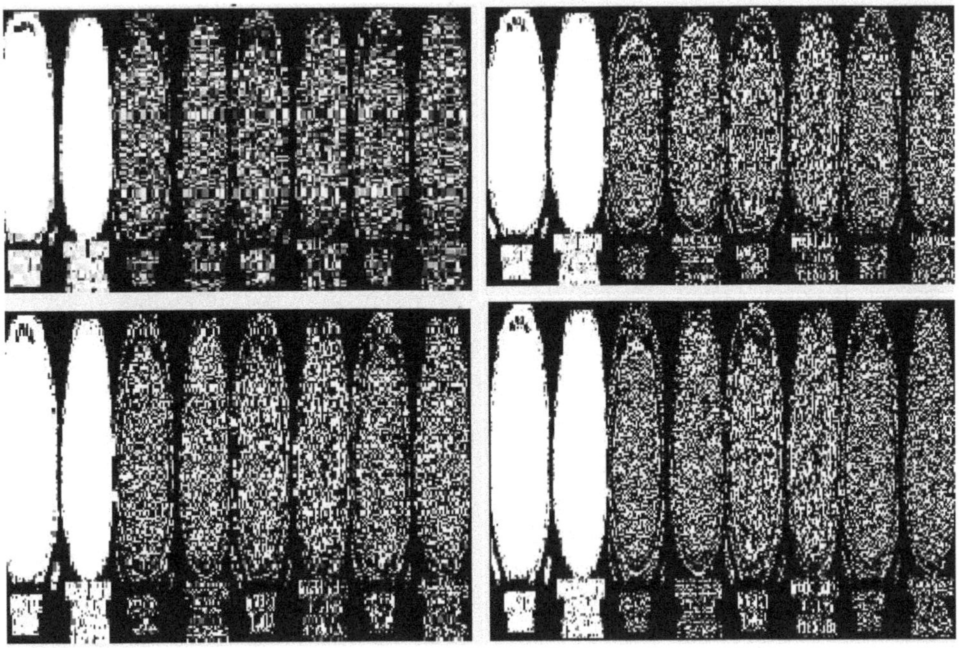

Figure 6. Compression of the transformed image.

As shown in Figure 7, the complete process of wavelet photograph compresion is carried out and explored as follows. The laptop takes an entered image, before which wavelet redecoration has been finished for the digital image, thresholding has been executed on the digital picture as shown in Figure 8, and entropy coding has been completed for the photograph, all of which are essential. Wavelet compression is a technique that lets one minimize the dimensions of files at the same time as the equal time, improving them via the elimination of high-frequency noise additives. The documents can, without problems, be decreased beneath 1% of their actual length. Wavelet radically change, used to look up a sign into high-quality frequency factors at incredible choice scales, allows a picture's spatial and frequency attributes to be concurrently revealed. In addition, competencies that could be undetected at one decision point might be easy to spot at another. One of the most vital blessings of wavelets is that they provide simultaneous localization in time and frequency areas. The most requirement in acquiring wavelets is that fast wavelet remodeling is computationally very speedy. Wavelets have the great advantage of being successful in isolating sign information. Lossless compression is more significant for programs where maintaining extraordinary files is essential. The drawback of this compression method is that it calls for massive archives to hold files after compression.

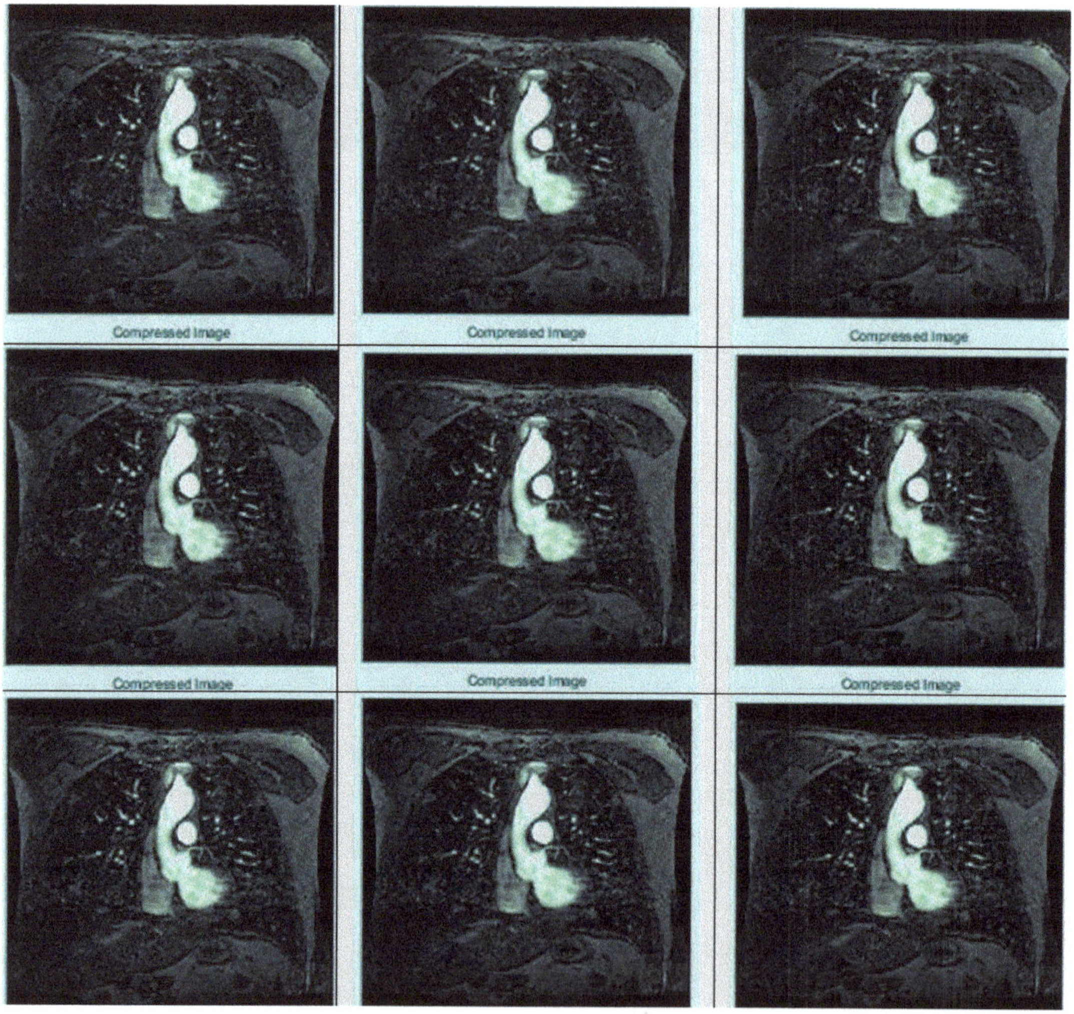

Figure 7. Multiple compression of MRI using different wavelet families.

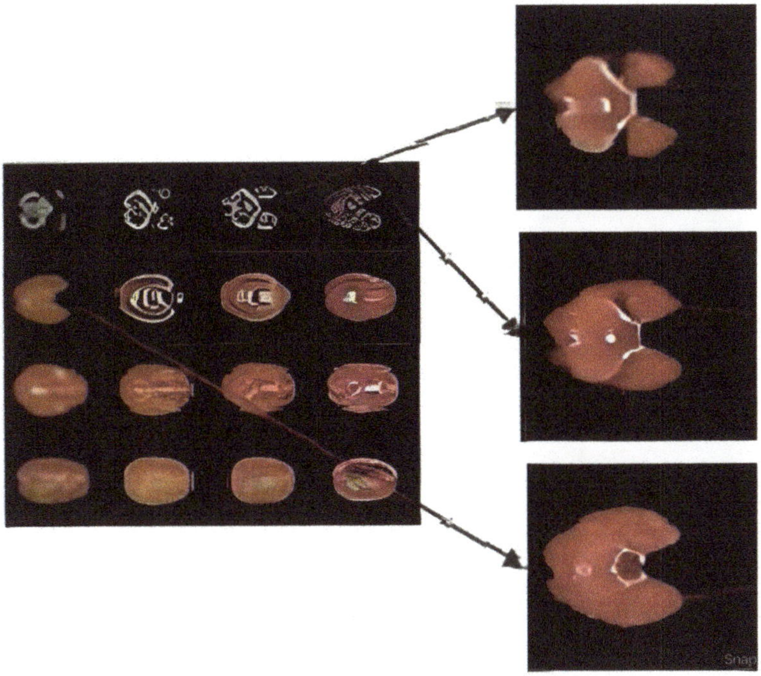

Figure 8. MRI compression using core wavelet families.

4.3. Comparative Analysis

The decompressed images of different sizes were compared with the original image. Several image measures—mean square error (MSE), signal-to-noise ratio (SNR), peak signal-to-noise ratio (PSNR), structural similarity index measure (SSIM), and bit error rate (BER)—were evaluated. The results are shown in Table 1. The performance of the proposed model was compared with those of other models—steganography, WT, segmentation, encryption and fusion, DCT, DWT, and classification [1–4,8,9,14–17,19,21,23,24].

Table 1. Comparative analysis.

Measures	Original Image	Decompressed Image 1	Decompressed Image 2
MSE	41.89	41.88	41.88
SNR	25.79	25.79	25.79
PSNR	31.92	31.91	31.91
SSIM	0.99	0.99	0.99
BER (number, ratio)	1548, 7.38%	1548, 7.37%	1548, 7.37%

This research proves that the evaluation parameters do not change significantly as the size of the information increases and gives the insight that, regardless of size, the error rate is low enough that the steganography or decompressed image will not be distorted. Regardless of changes to the input data, the embedding and compression algorithms do not deviate from expected results, proving robustness and accuracy to any dataset and the tricky challenges introduced. The compression ratios of all images also exceeded the 80% benchmark, demonstrating high compression ratios for lossless images. From the original 17.63 GB of the significant medical dataset, the total compression achieved was 8.62 GB, which is very efficient for medical repository storage. These proven results give a clear insight into how well the hybridization and fusion of these techniques have yielded results and give us a significant boost and inspiration to use this well-constructed technology.

4.4. Compression Analysis

The process of image steganography, transformation, and compression can be performed on any platform using any method. However, it is essential to have higher security and minimal storage space. Therefore, choosing an efficient and effective method will give the best results. Choosing a fast KT optimized for steganography and DWTs, along with wavelet packet compression of images, gives better results. To demonstrate the success of the proposed method, experiments were performed in MATLAB version R2017b, in which the data embedding, transformation, and compression work was carried out. MATLAB software is a tool that can be used to perform all mathematical operations, providing highly accurate results at every step. Windows 10 OS provides excellent support for productive and skillful research. KT lends an innovative flavor to steganography techniques, and transformations help protect steganography images. Additionally, the transformed image is compressed using lossless compression and decompressed again at the receiving end to the exact same resolution as the original image. The compressed images are stored in the database in less space, which helps in efficient storage.

4.5. Comparative Results and Statistical Analysis

The performance of the proposed method was also evaluated for various JPEG and PNG image formats to obtain compression ratios with percentages of compression. The image sizes considered for the simulation were 128 × 128, 256 × 256, 512 × 512, and 1024 × 1024. The performance measures of the proposed model are shown in Table 2. The expected performance in terms of compression ratios to other models is detailed in Table 3. The expected performance in terms of compression percentages to other models is shown in Table 4. The expected performance in terms of computing times for the proposed model to other models is shown in Table 5. The expected compression and decompression computing times for the benchmark datasets for the proposed model in relation to those of the other models are shown in Tables 6 and 7, respectively.

Table 2. JPEG and PNG format comparisons.

Size	Features	JPEG	PNG
128 × 128 Image 1	Compression ratio	8.9%	8.7%
128 × 128 Image 1	Compression %	52.5%	46.6%
128 × 128 Image 2	Compression ratio	8.5%	8.2%
128 × 128 Image 2	Compression %	53.5%	41.5%
256 × 256 Image 1	Compression ratio	8.1%	8.0%
256 × 256 Image 1	Compression %	42.5%	41.2%
256 × 256 Image 2	Compression ratio	7.5%	7.2%
256 × 256 Image 2	Compression %	36.2%	33.5%

Table 3. Expected performance—compression ratios compared with those of other models.

Size	Proposed Model	Other Models
128 × 128	8.6%	8.1–8.5%
256 × 256	8.4%	8.1–8.3%
512 × 512	7.9%	7.5–7.8%
1024 × 1024	7.8%	7.7–7.8%
Public datasets	8.5%	7.8–8.4%

Table 4. Expected performance—compression percentages compared with those of other models.

Size	Proposed Model	Other Models
128 × 128	58.6%	58.1–58.3%
256 × 256	48.4%	48.1–48.3%
512 × 512	57.2%	47.5–47.9%
1024 × 1024	37.8%	35.7–36.4%
Public datasets	58.4%	45.5–55.8%

Table 5. Expected performance—computing times (in ms) compared with those of other models.

Size	Proposed Model	Other Models
128 × 128	3.5	5.7
256 × 256	3.8	6.2
512 × 512	4.2	6.4
1024 × 1024	4.7	7.5
Public datasets	35.2	48.4

Table 6. Expected performance—compression computing times (in ms) compared with those of other models.

Size	Proposed Model	Other Models
128 × 128	2.7	3.8
256 × 256	3.2	5.9
512 × 512	4.1	5.4
1024 × 1024	4.2	5.5
Public datasets	12.56	21.78

Table 7. Expected performance—decompression computing times (in ms) compared with those of other models.

Size	Proposed Model	Other Models
128 × 128	3.8	4.7
256 × 256	4.3	5.2
512 × 512	4.3	6.2
1024 × 1024	4.5	7.8
Public datasets	15.23	25.35

The following significant inferences were drawn based on the simulation.

- JPEG formats' compression ratio percentages were slightly higher than PNG formats. When the image size increases, that is, for high-resolution images, the compression ratio lies between 7% and 7.5%, and the compression percentage lies between 30% and 37%.
- The proposed model increases the expected compression ratio and percentage compared to other models. The average compression ratio lies between 7.8% and 8.6%, and the expected compression ratio lies between 35% and 60%.
- Computing time is reduced with the proposed method relative to other methods. For high-resolution images, the expected computing time lies between 4 ms and 5 ms compared to other approaches.

The limitations of the proposed model are as follows:

- The memory requirement through DWT is higher since it processes the complete image.
- Training and inference can be achieved by designing an optimal neural network to reduce complexity.

5. Conclusions and Future Work

In this research, a picture of a brain hemorrhage was taken and entered into the patient's records. The input image was transformed using a DWT "branch" wavelet. The transformed image was then compressed using "Haar" wavelet compression. The image was compressed to 83.3333% and decompressed and reconstructed to obtain the embedded brace image. This downloaded brace image was then used to extract patient information from the original image. After decompression, the resolution of the studied image was never disturbed, and since the brace image was lossless, the extracted patient data were unchanged. The optimization obtained from the test significantly contributed to the preservation and long-term archiving of the medical image without compromising the patient's privacy. The compression technology produced an amazing lossless image after decompression, which was useful for retrieving patient information. The method developed in this research ensured that medical images were archived optimally and increased privacy. JPEG formats' compression ratio percentages were slightly higher than those of PNG formats. When image size increases, that is, for high-resolution images, the compression ratio lies between 7% and 7.5%, and the compression percentage lies between 30% and 37%. The proposed model increases the expected compression ratio and percentage compared to other models. The average compression ratio lies between 7.8% and 8.6%, and the expected compression ratio lies between 35% and 60%. Computing time is reduced with the proposed method relative to other methods. For high-resolution images, the expected computing time lies between 4 ms and 5 ms.

In the future, the Internet of Medical Things, medical imaging strategies, soft computing with evolutionary operators, and other hybrid image processing strategies will be applied further to reduce the complexity of the proposed model and to develop better recommender systems for large-scale, high-quality medical images.

Author Contributions: Conceptualization, X.X., R.R. (Rengasri Rajan), O.I.K. and R.M.; methodology, S.K.R.; software, S.K.R., G.M.A.; validation, R.M., S.K.R. and R.R. (Rangarajan Raghavan); formal analysis, R.M.; investigation, R.R. (Rangarajan Raghavan), G.M.A.; resources, S.K.R.; data curation, X.X., G.M.A.; writing—original draft preparation, G.M.A.; writing—review and editing, R.R. (Rengasri Rajan); visualization, O.I.K.; supervision, R.M., R.R. (Rengasri Rajan); project administration, X.X.; funding acquisition, O.I.K., G.M.A. All authors have read and agreed to the published version of the manuscript.

Funding: This work was supported by the National Natural Science Foundation of China (no. 62172095) and the Natural Science Foundation of Fujian Province (nos. 2020J01875 and 2022J01644).

Institutional Review Board Statement: Not applicable.

Informed Consent Statement: Not applicable.

Data Availability Statement: Not available.

Conflicts of Interest: The authors declare no conflict of interest.

References

1. Cheddad, A.; Condell, J.; Curran, K.; Mc Kevitt, P. Digital image steganography: Survey and analysis of current methods. *Signal Process.* **2010**, *90*, 727–752. [CrossRef]
2. Averbuch, A.; Lazar, D.; Israeli, M. Image compression using wavelet transform and multiresolution decomposition. *IEEE Trans. Image Process.* **1996**, *5*, 4–15. [CrossRef] [PubMed]
3. Bhadauria, N.S.; Bist, M.S.; Patel, R.B.; Bhadauria, H.S. Performance evaluation of segmentation methods for brain CT images based hemorrhage detection. In Proceedings of the 2015 2nd International Conference on Computing for Sustainable Global Development (INDIACom), New Delhi, India, 11–13 March 2015.
4. Brahimi, T.; Laouir, F.; Boubchir, L.; Ali-Chérif, A. An improved wavelet-based image coder for embedded greyscale and color image compression. *AEU-Int. J. Electron. Commun.* **2017**, *73*, 183–192. [CrossRef]
5. Bruylants, T.; Munteanu, A.; Schelkens, P. Wavelet based volumetric medical image compression. *Signal Process. Image Commun.* **2015**, *31*, 112–133. [CrossRef]

6. Reegu, F.A.; Abas, H.; Jabbari, A.; Akmam, R.; Uddin, M.; Wu, C.-M.; Chen, C.-L.; Khalaf, O.I. Interoperability Requirements for Blockchain-Enabled Electronic Health Records in Healthcare: A Systematic Review and Open Research Challenges. *Secur. Commun. Netw.* **2022**, *2022*, 9227343. [CrossRef]
7. Chang, C.-C.; Tseng, H.-W. A steganographic method for digital images using side match. *Pattern Recognit. Lett.* **2004**, *25*, 1431–1437. [CrossRef]
8. Datta, B.; Mukherjee, U.; Bandyopadhyay, S.K. LSB Layer Independent Robust Steganography using Binary Addition. *Procedia Comput. Sci.* **2016**, *85*, 425–432. [CrossRef]
9. Ganesan, K.; Rajaguru, H. Performance Analysis of Wavelet Function Using Denoising for Clinical Database. In *Intelligent and Efficient Electrical Systems*; Springer: Singapore, 2018; pp. 241–249. [CrossRef]
10. Jung, K.-H.; Ha, K.-J.; Yoo, K.-Y. Image Data Hiding Method Based on Multi-Pixel Differencing and LSB Substitution Methods. In Proceedings of the 2008 International Conference on Convergence and Hybrid Information Technology, Daejeon, Korea, 28–30 August 2008; pp. 355–358. [CrossRef]
11. Karthikeyan, C.; Palanisamy, C. An Efficient Image Compression Method by Using Optimized Discrete Wavelet Transform and Huffman Encoder. *J. Comput. Theor. Nanosci.* **2018**, *15*, 289–298. [CrossRef]
12. Suryanarayana, G.; Chandran, K.; Khalaf, O.I.; Alotaibi, Y.; Alsufyani, A.; Alghamdi, S.A. Accurate Magnetic Resonance Image Super-Resolution Using Deep Networks and Gaussian Filtering in the Stationary Wavelet Domain. *IEEE Access* **2021**, *9*, 71406–71417. [CrossRef]
13. Banumathy, D.; Khalaf, O.I.; Romero, C.A.T.; Indra, J.; Sharma, D.K. CAD of BCD from Thermal Mammogram Images Using Machine Learning. *Intell. Autom. Soft Comput.* **2022**, *34*, 595–612. [CrossRef]
14. Liu, X.; Mei, W.; Du, H. Simultaneous image compression, fusion and encryption algorithm based on compressive sensing and chaos. *Opt. Commun.* **2016**, *366*, 22–32. [CrossRef]
15. Mehto, A.; Mehra, N. Adaptive Lossless Medical Image Watermarking Algorithm Based on DCT & DWT. *Procedia Comput. Sci.* **2016**, *78*, 88–94. [CrossRef]
16. Nambutdee, A.; Airphaiboon, S. Medical image encryption based on DCT-DWT domain combining 2D-DataMatrix Barcode. In Proceedings of the 2015 8th Biomedical Engineering International Conference (BMEiCON), Pattaya, Thailand, 25–27 November 2015; pp. 1–5.
17. Nayak, D.R.; Dash, R.; Majhi, B. Brain MR image classification using two-dimensional discrete wavelet transform and AdaBoost with random forests. *Neurocomputing* **2016**, *177*, 188–197. [CrossRef]
18. Snehkunj, R.; Jani, A.N. A systematic approach for brain abnormality identification from biomedical images. In Proceedings of the 2016 International Conference on Data Mining and Advanced Computing (SAPIENCE), Ernakulam, India, 16–18 March 2016; pp. 210–213. [CrossRef]
19. Talukder, K.H.; Harada, K. Haar wavelet based approach for image compression and quality assessment of compressed image. *arXiv* **2010**, arXiv:1010.4084.
20. Tavera, C.A.; Ortiz, J.H.; Khalaf, O.I.; Saavedra, D.F.; Aldhyani, T.H. Wearable wireless body area networks for medical applications. *Comput. Math. Methods Med.* **2021**, *2021*, 5574376. [CrossRef]
21. Sudhakar, S.; Khalaf, O.I.; Sharma, D.K.; Hamad, A.A. Secured and privacy-based IDS for healthcare systems on E-medical data using machine learning approach. *Int. J. Reliab. Qual. E-Healthc. (IJRQEH)* **2022**, *11*, 1–11.
22. Sengan, S.; Khalaf, O.I.; Rao, G.R.K.; Sharma, D.K.; Amarendra, K.; Hamad, A.A. Security-aware routing on wireless communication for E-health records monitoring using machine learning. *Int. J. Reliab. Qual. E-Healthc. (IJRQEH)* **2022**, *11*, 1–10. Available online: https://econpapers.repec.org/article/iggjrqeh0/v_3a11_3ay_3a2022_3ai_3a3_3ap_3a1-10.htm (accessed on 7 December 2022). [CrossRef]
23. Wang, X.; Peng, Y.; Lu, L.; Lu, Z.; Bagheri, M.; Summers, R.M. Chestx-ray8: Hospital-scale chest X-ray database and benchmarks on weakly-supervised classification and localization of common thorax diseases. In Proceedings of the IEEE Conference on Computer Vision and Pattern Recognition (CVPR), Honolulu, HI, USA, 21–26 July 2017; pp. 3462–3471.
24. Wu, H.-C.; Wu, N.-I.; Tsai, C.-S.; Hwang, M.-S. Image steganographic scheme based on pixel-value differencing and LSB replacement methods. *IEEE Proc. Vis. Image Signal Process.* **2005**, *152*, 611–615. [CrossRef]
25. Marappan, R.; Sethumadhavan, G. Solution to Graph Coloring Using Genetic and Tabu Search Procedures. *Arab. J. Sci. Eng.* **2017**, *43*, 525–542. [CrossRef]
26. Marappan, R.; Sethumadhavan, G. Complexity Analysis and Stochastic Convergence of Some Well-known Evolutionary Operators for Solving Graph Coloring Problem. *Mathematics* **2020**, *8*, 303. [CrossRef]
27. Marappan, R.; Sethumadhavan, G. Solving Graph Coloring Problem Using Divide and Conquer-Based Turbulent Particle Swarm Optimization. *Arab. J. Sci. Eng.* **2021**, *47*, 9695–9712. [CrossRef]
28. Bhaskaran, S.; Marappan, R. Design and analysis of an efficient machine learning based hybrid recommendation system with enhanced density-based spatial clustering for digital e-learning applications. *Complex Intell. Syst.* **2021**, *1*, 1–17. [CrossRef]
29. Marappan, R.; Sethumadhavan, G. A New Genetic Algorithm for Graph Coloring. In Proceedings of the 2013 Fifth International Conference on Computational Intelligence, Modelling and Simulation, Seoul, Korea, 24–25 September 2013; pp. 49–54. [CrossRef]
30. Sethumadhavan, G.; Marappan, R. A genetic algorithm for graph coloring using single parent conflict gene crossover and mutation with conflict gene removal procedure. In Proceedings of the 2013 IEEE International Conference on Computational Intelligence and Computing Research, Enathi, India, 26–28 December 2013; pp. 1–6. [CrossRef]

31. Marappan, R.; Sethumadhavan, G. Divide and conquer based genetic method for solving channel allocation. In Proceedings of the 2016 International Conference on Information Communication and Embedded Systems (ICICES), Chennai, India, 25–26 February 2016; pp. 1–5. [CrossRef]
32. Ali, Y.H.; Chinnaperumal, S.; Marappan, R.; Raju, S.K.; Sadiq, A.T.; Farhan, A.K.; Srinivasan, P. Multi-Layered Non-Local Bayes Model for Lung Cancer Early Diagnosis Prediction with the Internet of Medical Things. *Bioengineering* **2023**, *10*, 138. [CrossRef]
33. Rahman, A.; Hamada, M. Lossless Image Compression Techniques: A State-of-the-Art Survey. *Symmetry* **2019**, *11*, 1274. [CrossRef]
34. Yamagiwa, S.; Yang, W.; Wada, K. Adaptive Lossless Image Data Compression Method Inferring Data Entropy by Applying Deep Neural Network. *Electronics* **2022**, *11*, 504. [CrossRef]

bioengineering MDPI

Article

On the Bioconvective Aspect of Viscoelastic Micropolar Nanofluid Referring to Variable Thermal Conductivity and Thermo-Diffusion Characteristics

Omar T. Bafakeh, Kamel Al-Khaled, Sami Ullah Khan, Aamar Abbasi, Charankumar Ganteda and M. Ijaz Khan et al.

1 Department of Industrial Engineering, Jazan University, Jazan 82822, Saudi Arabia
2 Department of Mathematics & Statistics, Jordan University of Science and Technology, P.O. Box 3030, Irbid 22110, Jordan
3 Department of Mathematics, COMSATS University Islamabad, Sahiwal 57000, Pakistan
4 Department of Mathematics, University of Azad Jammu and Kashmir, Muzaffarabad 13100, Pakistan
5 Department of Engineering Mathematics, College of Engineering, Koneru Lakshmaiah Education Foundation, Vaddeswaram 522302, Andhra Pradesh, India
6 Department of Mathematics and Statistics, Riphah International University I-14, Islamabad 44000, Pakistan
7 Department of Mechanical Engineering, Lebanese American University, Beirut 1102 2801, Lebanon
8 Mechanical Engineering Department, College of Engineering and Islamic Architecture, Umm Al-Qura University, P.O. Box 5555, Makkah 21955, Saudi Arabia
9 Center of Research, Faculty of Engineering, Future University in Egypt, New Cairo 11835, Egypt
* Correspondence: scientificresearchglobe@gmail.com or ijazfmg_khan@yahoo.com

Abstract: The bioconvective flow of non-Newtonian fluid induced by a stretched surface under the aspects of combined magnetic and porous medium effects is the main focus of the current investigation. Unlike traditional aspects, here the viscoelastic behavior has been examined by a combination of both micropolar and second grade fluid. Further thermophoresis, Brownian motion and thermodiffusion aspects, along with variable thermal conductivity, have also been utilized for the boundary process. The solution of the nonlinear fundamental flow problem is figured out via convergent approach via Mathematica software. It is noted that this flow model is based on theoretical flow assumptions instead of any experimental data. The efficiency of the simulated solution has been determined by comparing with previously reported results. The engineering parameters' effects are computationally evaluated for some definite range.

Keywords: bioconvection flow; viscoelastic micropolar fluid; heat transfer; double diffusion; variable thermal conductivity

Citation: Omar T. Bafakeh, Kamel Al-Khaled, Sami Ullah Khan, Aamar Abbasi, Charankumar Ganteda and M. Ijaz Khan et al. On the Bioconvective Aspect of Viscoelastic Micropolar Nanofluid Referring to Variable Thermal Conductivity and Thermo-Diffusion Characteristics. *Bioengineering* **2023**, *10*, 73. https://doi.org/10.3390/bioengineering10010073

Academic Editors: Roderick Melnik, Sundeep Singh and Esther Pueyo

Received: 20 October 2022
Revised: 16 December 2022
Accepted: 30 December 2022
Published: 5 January 2023

1. Introduction

At present, noteworthy attention is being paid to the study of nanofluids, due to their advanced warm transportation features. Experimental-based investigations report that the thermal performances of nanoparticles may intestinally affect the shape and size of particles, the volume fraction of nanoparticles, the material of particles and traditional base liquids. Several prospective applications of nanofluids related to chemical, biological, mechanical and engineering areas may include cooling of laptop processors, fuel chambers, enhancement of chemical reactions, air purifiers, fission reactions, turbine engines, diagnosis and treatment of tumors, damaging risky tissues in the human body and improving thermal efficiency of raw petroleum materials. Such nanoparticles, originally proposed by Choi, are in the micro-size range (1–100 nm) [1]. Based on the excrementally based observations of [1], nanoparticles have been recommended as the most convenient, cheap and environmentally friendly source for improving thermal conductivity, as comparison to conventional base liquids. Due to such illustrious features, a variety of contributions have been listed by numerous researchers. Buongiorno [2] advised on the most interesting slip features and thermo-physical aspects of nanoparticles, namely thermophoresis and Brownian motion, by developing a non-homogeneous model for the transportation equations. Reddy et al. [3] constituted the slip effects in an unsteady MHD flow of nanofluid along with thermophoresis and Brownian movement features endorsed by a slendering surface. Usman et al. [4] derived a numerical solution for the movement of Casson fluid immersed in nanoparticles with slip effects. The desired flow was objected over a cylinder and the treated numerical method was a collocation technique. Another attempt, focusing on the transportation of disk flow of nanomaterial was predicted by Yin et al. [5]. Madhu et al. [6] performed a theoretical simulation for unsteady Maxwell nanofluid flow via a stretched surface. Turkyilmazoglu [7] initiated the transportation mechanism nanofluid extrusion for free as well as circular jets, and governed equations were treated exactly. We acknowledge another continuation of Turkyilmazoglu [8] regarding the channel flow of nanofluid by using the famous Buongiorno model. Interestingly, the results were also predicted for the pure fluid case and results were reported via the integration technique. Abbasi et al. [9] visualized the Hall mechanism of nanofluid with peristaltic movement under the object of entropy generation. The microorganisms of nanofluid with behavior of slip effects were depicted by Waqas et al. [10]. Some more research on this topic can be seen in [11–15].

It is commonly noted that various transportation systems involve simultaneous heat and mass transfer effects in a moving fluid. In such circumstances, the flux is intricate with respect to both temperature and concentration gradients and notifies as thermodiffusion effects. Such effects may attain prime importance in cases where flow regime governs the different densities. The double diffusion convection phenomenon has key applications in the evaluation of systems which happen with various densities. Agarwal et al. [15] performed investigations for diffusion convective flow of nanofluid in a rotating frame. Raghunatha et al. [16] proved the thermal stability of viscoelastic material for triple diffusion phenomena. Daba et al. [17] discussed the injection onset for tripe diffusive flow in vertical surfaces. Gireesha et al. [18] observed the 3D analysis for tripe diffusion flow numerically. The numerical approach for Prandtl nanofluid flow under double diffusion phenomena was justified by Bilal et al. [19]. Hayat et al. [20] used Cattaneo–Christov relation for double diffusion phenomena with Jeffrey material. Irfan et al. [21] used the variable thermal impact for double diffusion flow. The Soret effects for Maxwell material in a bidirectional regime was investigated by Prasannakumara et al. [22].

The analogous characteristics of non-Newtonian fluids have intrigued engineers and scientist recently owing to their peculiar rheological properties. Various industrial and biological applications have been found for non-Newtonian fluid paints, petroleum products, fiber coating, food industries, engine oils, glass blowing, blood, pharmaceuticals, etc. Due to the diverse features of nonlinear liquids, research is undertaken by scientists in different manners and indicates different mathematical relations for each non-Newtonian model. Listing such rheological models, micropolar fluid is one which explains the behavior of materials under the assessment of a micro-level approach and on a micro basis. The justifications of mathematical expressions for linear as well as couple models can be retained as a limiting case. The basic concept of micropolar fluid was deduced by Eringen [23–26] and later on many successful contributions were carried out by numerous scientists. For instance, Kabeir [27] examined the stagnation point flow of a micropolar fluid model under the action of a magnetic field. Nazeer et al. [27] and Abbasi et al. [28] worked on heat transport in fluid flow over a diverse geometries. Turkyilmazoglu [29] established some interesting exact solutions for micropolar fluid induced by a flat plate. Sui et al. [30] utilized nonlinear diffusive flow of micropolar fluid over a stretched surface. Sajid et al. [31] numerically analyzed the stagnation point flow over a spiraling disk.

Owing to the above-mentioned chemical, mechanical and engineering significance, the aim here is to present a study of double diffusion flow of viscoelastic micropolar fluid by appending magnetized nanoparticles containing the microorganisms. The viscoelastic micropolar fluid refers to the decomposition of second grade fluid and micropolar fluid refers to the generalized viscoelastic model. The present analysis extends the work of [6] by considering the bioconvection phenomenon, combining magnetic and porous medium effects, interacting the nanoparticles and interpolating double diffusion effects. The analytical expressions which govern the solution of the modeled problem are conveyed through the most pragmatic convergent method, namely the homotopy analysis technique. The realistic physical description and significance are evaluated for each physical parameter. These results can be attributed to the enhancement of thermal efficiency in engineering devices. Further, this model enables us to present simultaneous results for viscous, viscoelastic, micropolar and couple stress fluid.

2. Flow Problem

Current continuation reports the bioconvective determination of viscoelastic micropolar nanofluid under the influence of a magnetic field and porous medium. The flow model is based on theoretical flow assumptions. The Buongiorno model is incorporated to see the significance of Brownian motion and thermophoresis impact. We originate a coordinate system from the edge of a stretched surface where velocity component u is considered in the x-direction while the v component is utilized in the y-direction. The magnetic field effects are targeted vertically. Further, the temperature, solutal nanofluid concentration and volume fraction are denoted by $(T,\ C,\ \varphi)$. The ambient fluid is retracted by velocity $u = bx$, where b is constant. Keeping these assumptions in mind, the associated model is explicated as [27,32]:

$$\frac{\partial u}{\partial x} + \frac{\partial v}{\partial y} = 0, \tag{1}$$

$$u\frac{\partial u}{\partial x} + v\frac{\partial u}{\partial y} = \frac{k_1}{\rho_f}\frac{\partial N}{\partial y} + \left(\nu + \frac{k_1}{\rho_f}\right)\frac{\partial^2 u}{\partial y^2} - k_0\left[u\frac{\partial^3 u}{\partial x \partial y^2} + \frac{\partial u}{\partial x}\frac{\partial^2 u}{\partial y^2} - \frac{\partial u}{\partial y}\frac{\partial^2 u}{\partial x \partial y^2} + v\frac{\partial^3 u}{\partial x^3}\right] - \left(\frac{\sigma_e B_0^2}{\rho_f} + \frac{\nu\vartheta}{k^*}\right)u, \tag{2}$$

$$u\frac{\partial N}{\partial x} + v\frac{\partial N}{\partial y} = \frac{\gamma}{\rho_f j}\frac{\partial^2 N}{\partial y^2} - \frac{k_1}{\rho_f j}\left(2N + \frac{\partial u}{\partial y}\right), \tag{3}$$

$$u\frac{\partial T}{\partial x} + v\frac{\partial T}{\partial y} = \frac{1}{(\rho c_p)_f}\frac{\partial}{\partial y}\left(K(T)\frac{\partial T}{\partial y}\right) + \tau_T\left[D_T\frac{\partial \phi}{\partial y}\frac{\partial T}{\partial y} + \frac{D_T}{T_\infty}\left(\frac{\partial T}{\partial y}\right)^2\right] + DK_{TC}\frac{\partial^2 C}{\partial y^2}, \tag{4}$$

$$u\frac{\partial C}{\partial x} + v\frac{\partial C}{\partial y} = D_s\frac{\partial^2 C}{\partial y^2} + DK_{CT}\frac{\partial^2 T}{\partial y^2}, \tag{5}$$

$$u\frac{\partial \varphi}{\partial x} + v\frac{\partial \varphi}{\partial y} = D_B\frac{\partial^2 \varphi}{\partial y^2} + \frac{D_T}{T_\infty}\frac{\partial^2 T}{\partial y^2}, \tag{6}$$

$$u\frac{\partial n}{\partial x} + v\frac{\partial n}{\partial y} + \frac{bW_c}{(C_w - C_\infty)}\left[\frac{\partial}{\partial y}\left(n\frac{\partial C}{\partial y}\right)\right] = D_m\left(\frac{\partial^2 n}{\partial y^2}\right), \tag{7}$$

Here k_1 determines the fluid constant, k_0 denotes the viscoelastic parameter, σ_e signifies the electrical conductivity, ϑ stands for the permeability of porous medium, γ spin gradient viscosity, ρ_f fluid density, j symbolizes the microinertia per unit mass, N relates the microrotation, α_1 stands for thermal diffusivity, D_B denotes the Brownian diffusion coefficients, DK_{TC} is Dufour diffusivity, D_s determine solutal diffusivity, D_T notify the thermophoretic diffusion coefficient, DK_{CT} denote Soret diffusivity, $\tau_T = (\rho c)_p/(\rho c)_f$ highlights the heat capacity of nanoparticles to heat capacity of fluid ratio.

Current analysis has been performed by articulating the following boundary conditions

$$u = u_\omega = bx,\ \ v = 0,\ \ N = 0,\ \ T = T_w,\ \ C = C_w, \varphi = \varphi_w, n = n_w \text{ at } y = 0, \tag{8}$$

$$u \to 0,\ \frac{\partial u}{\partial y} \to 0, u \to 0,\ \ N \to 0, T \to T_\infty,\ \ C \to C_\infty, \varphi \to \varphi_\infty, n \to n_\infty \text{ at } y \to \infty. \tag{9}$$

For variable conductivity, Equation (4) may be amended by introducing

$$K(T) = K_{\infty}\left[1 + \varepsilon\frac{(T - T_{\infty})}{\Delta T}\right], \tag{10}$$

with K_{∞} (material conductivity) and ε (thermal dependence conductivity factor).

Let us insinuate the appropriate variables to attain the non-dimensional forms [33]

$$u = bxf'(\eta),\ v = -\sqrt{bx}f(\eta), N = \sqrt{\frac{b}{\nu}}bxg(\eta),\ \eta = \sqrt{\frac{a}{\nu}}y, \tag{11}$$

$$\theta(\eta) = \frac{T - T_{\infty}}{T_w - T_{\infty}}, s(\eta) = \frac{C - C_{\infty}}{C_w - C_{\infty}}, \phi(\eta) = \frac{\varphi - \varphi_{\infty}}{\varphi_w - \varphi_{\infty}}, \chi(\eta) = \frac{n - n_{\infty}}{n_w - n_{\infty}}. \tag{12}$$

After invoking above quantities, Equations (3)–(5) are re-established into following forms

$$(1 + K)f''' + Kg' + ff'' - \left(f'\right)^2 - Haf' - k\left[2f'f''' - (f'')^2 - ff''''\right] = 0, \tag{13}$$

$$\left(1 + \frac{K}{2}\right)g'' - K(2g + f'') - f'g + g'f = 0, \tag{14}$$

$$(1 + \varepsilon\theta)\theta'' + \varepsilon\left(\theta'\right)^2 + \Pr\left[f\phi' + Nb\theta'\phi' + Nt\left(\theta'\right)^2 + (Nd)s''\right] = 0, \tag{15}$$

$$s'' + Le\left(f\phi'\right) + Ld\theta_{yy} = 0, \tag{16}$$

$$\phi'' + Ln\left(f\phi'\right) + \frac{Nt}{Nb}\theta'' = 0, \tag{17}$$

$$\chi'' + Lbf\chi' - Pe\left(\phi''(\chi + \delta_1) + \chi'\phi'\right) = 0. \tag{18}$$

Governing dimensionless boundary conditions are

$$f'(0) = 1,\ f(0) = 0, g(0) = 0,\ \theta(0) = 1,\ s(0) = 1, \phi(0) = 1, \chi(0) = 1, \tag{19}$$

$$f'(\infty) \to 0, f''(\infty) \to 0,\ g(\infty) \to 0, \theta(\infty) \to 0, s(\infty) \to 0, \phi(\infty) \to 0, \chi(\infty) \to 0. \tag{20}$$

We define the most important engineering parameters, such as viscoelastic parameter k, combined parameter H_a, vortex viscosity parameter K, Prandtl number Pr,, thermophoresis parameter Nt, Brownian motion parameter Nb, modified Dufour number Nd, regular Lewis number Le, Dufour Lewis number Ld, Peclet number Pe, bioconvection Lewis Lb, microorganism concentration difference δ_1 and nano Lewis number Ln as follows

$$\begin{aligned}
&k = bk_0/\nu\rho_f,\\
&Ha = \sqrt{\sigma_e B_0^2/\rho_f b + \nu\vartheta/bk^*},\\
&K = k_1/\rho_f\nu,\\
&\Pr = \nu/\alpha_f,\\
&N_t = (\rho c)_p D_T(T_w - T_\infty)/(\rho c)_f T_\infty\nu, N_b = (\rho c)_p D_B(C_w - C_\infty)/(\rho c)_f\nu,\\
&Sc = \nu/D_m,\\
&Le = \nu/D_s,\\
&Ld = D_{CT}(T_w - T_\infty)/\alpha_m(C_w - C_\infty), Lb = \tfrac{\nu}{D_m}, Pe = \tfrac{bw_c}{D_m}, Ln = \nu/D_B, \delta_1 = \tfrac{n_\infty}{n_w - n_\infty},\\
&Nd = D_{TC}(C_w - C_\infty)/\alpha_m(T_w - T_\infty).
\end{aligned}$$

To illustrates the physical features of shear stress, one must determine the skin friction coefficient in following, which is of flowing form

$$C_f = \frac{\tau_w}{\rho u_w^2}, \tau_w = \left[(\mu + k_1)\frac{\partial u}{\partial y} + k_1 N\right]_{y=0}, C_f\sqrt{\mathrm{Re}_x} = (1 + K)f''(0). \quad (21)$$

Moreover, the relation for wall couple stress is actuated as

$$M_x = \frac{\gamma\left(\frac{\partial N}{\partial y}\right)_{y=0}}{\rho b^2 x^3}, M_x\sqrt{\mathrm{Re}_x} = KGg'(0). \quad (22)$$

The definitions of the dimensionless local Nusselt number, local Sherwood number and local nano Sherwood number results in the following expressions

$$Nu_x\mathrm{Re}_x^{-1/2} = -\theta'(0),\ Sh_x\mathrm{Re}_x^{-1/2} = -s'(0),\ Sh_n\mathrm{Re}_x^{-1/2} = -\phi'(0), \mathrm{Re}_x^{-1/2}N_n = -\chi'(0). \quad (23)$$

where $\mathrm{Re}_x = u_w\overline{x}/v$ is mentioned for the local Reynolds number.

3. Homotopy Analysis Method

The nonlinear problems play essentially decisive roles in various disciplines of science, applied engineering and mathematics. Such exciting nonlinearity compels scientists to compose analytical or numerical approaches. Taking into account such analytical methods, the homotopy analysis method (HAM) technique is noted as the preferred dynamical technique due to its well confined accuracy. Unlike the perturbation method, no restriction is noticed for small or larger flow factors. This renewed technique was formerly advocated by Liao [34], and subsequently many equations based on mathematical and interdisciplinary sciences have been successfully treated by introducing this method. Since this technique is familiar from recent years, the details of this method have been omitted here. Readers are referred to some continuations in the literature [33,35–37].

4. Convergent Region and Solution Confirmations

The classical homotopic solution contains some interesting aspects, namely h_f, h_g, h_θ, h_ϕ and h_s in which are confined some specified regions, to examine the convergence of the HAM method. Obviously, apposite selection of these parameters conferred accurate convergence path and can be determined by $h-$ curves. Figure 1 utilizes such $h-$ curves for $f''(0), g'(0), \theta'(0), s'(0)$ and $\phi'(0), \chi'(0)$. Based on observations, the convergence regions for $f''(0), g'(0), \theta'(0), s'(0), \phi'(0)$ and $\chi'(0)$ have been listed in Table 1.

Table 1. Numerical evaluation for convergence region for h_f, h_g, h_θ, h_s, and h_ϕ.

Approximation Solutions	Auxiliary Parameters	Convergence Region
f	h_f	$-1.8 \le h_f \le -0.1$
g	h_g	$-1.8 \le h_g \le 0$
θ	h_θ	$-1.2 \le h_\theta \le -0.1$
s	h_s	$-1.3 \le h_s \le -0.2$
ϕ	h_ϕ	$-1.7 \le h_\phi \le -0.2$
χ	h_χ	$-1.5 \le h_\chi \le 0.0$

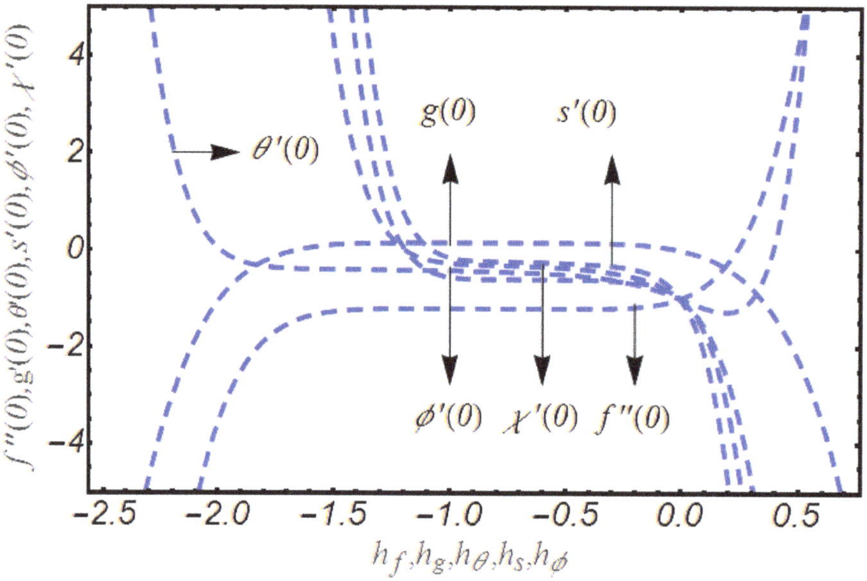

Figure 1. Curves for $f''(0)$, $g'(0)$, $\theta'(0)$, $s'(0)$, $\phi'(0)$ and $\chi'(0)$.

5. Verification of Results

The results are validated in Table 2 for the limiting case with the investigation of El-Kabeir [38]. The claimed results show fine accuracy with these available numerical results.

Table 2. Results verification for $f''(0)$ when $\alpha = 0$.

Ha	El-Kabeir [38]	Present Results
0.2	1.24857	1.24856
0.4	1.29537	1.29535
0.6	1.36988	1.36980

6. Analysis of Results

In this section, graphical results for the involved parameters are presented for $k = 0.0$ and $k = 0.5$.

It is mentioned that in order to vary the flow, parameters have been assigned constant values like $K = 0.5$, $Ha = 0.2$, $\varepsilon = 0.5$, $Nd = 0.2$, $\Pr = 0.7$, $Lb = 0.3$, $Pe = 0.2$, $Le = 0.3$, $Ld = 0.3$ and $Ln = 0.3$.

The effects of vortex viscosity factor K, variable thermal conductivity parameter ε, thermophoresis constant Nt, modified Dufour number Nd, Prandtl number Pr, and viscoelastic parameter k on temperature profile θ are visualized by plotting Figure 2a–f. Here, it is remarked that the physical observations for each parameter have been examined for viscoelastic behavior $(k = 0.5)$ and without the viscoelastic parameter i.e., $k = 0.0$. First, the variation of K is listed in Figure 2a. The temperature profile decreases by increasing K. In Figure 2b, we conclude that the temperature of nanoparticles for both $k = 0.0$ and $k = 0.5$ declines by varying Pr. This declining trend is relatively dominant for the viscoelastic case $k = 0.5$ as compared to $k = 0.0$. Physically, smaller physical attribution of thermal diffusivity for increasing Pr is noticed, which depressed the temperature. The response of variable thermal conductivity ε on θ is reported in Figure 2c. Following the graphical trend, it is concluded that variable thermal conductivity is more useful to enhance the temperature profile. Further, in contrast to $k = 0.0$, the increasing trend seems to be dominant for $k = 0.5$.

From Figure 2d, θ gradually increases when Nt is enlarged, that is, due to thermophoresis, the tiny fluid particles shift to the relatively cold region, with a resulting enhancement of temperature. Moreover, the thermal boundary layer is thicker for $k = 0.5$ than $k = 0.0$. The output of θ for modified Dufour number Nd is reported in Figure 2e. The variation of Nd leads to improving the temperature profile and enlarge the thickness of boundary layer. The observations for θ against various values of k are provided in Figure 2f. The reflection of viscoelastic factor reduced thermal flow due to presence of viscosity for both $K = 0.0$ and $K = 0.5$.

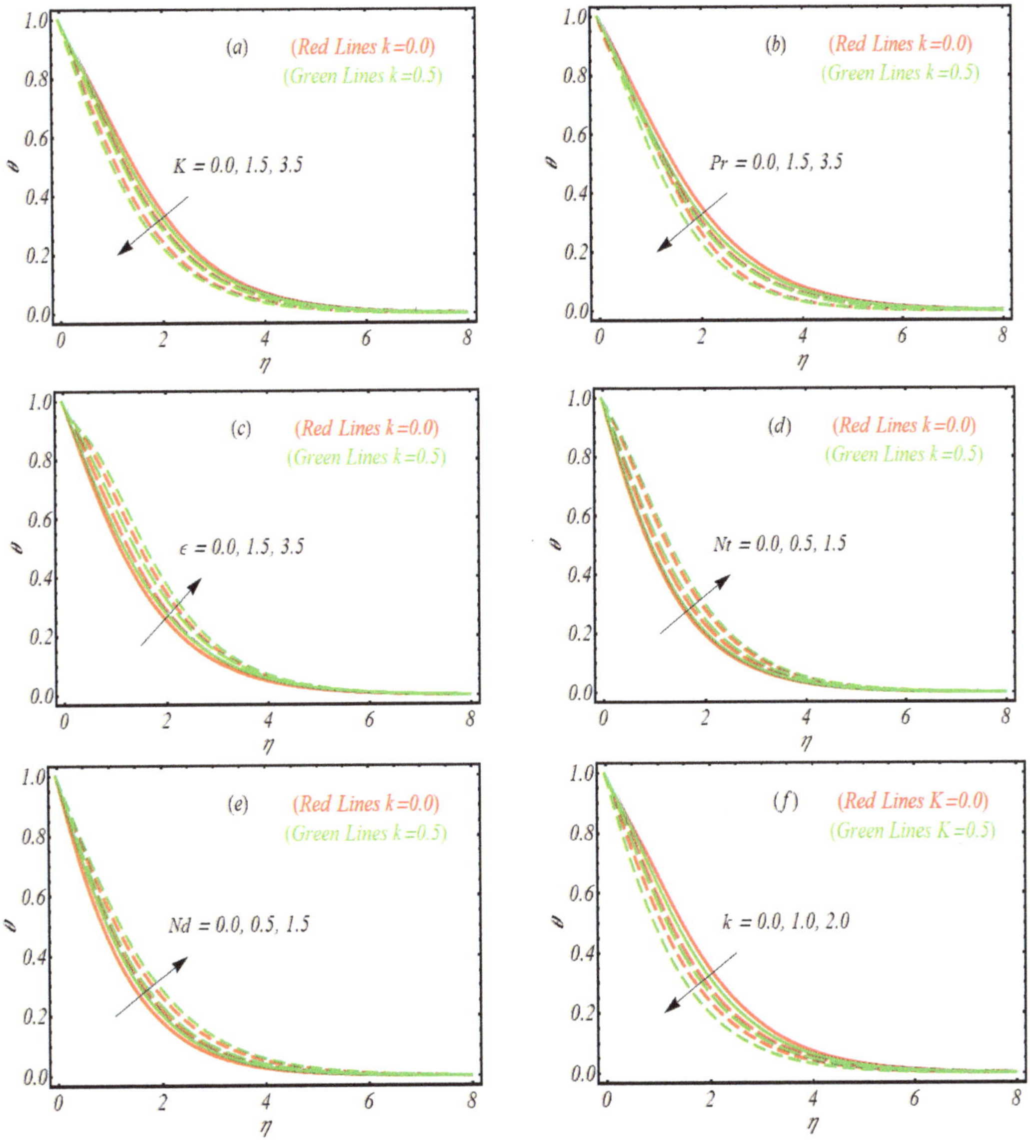

Figure 2. Temperature profile for (**a**) K, (**b**) Pr, (**c**) ε, (**d**) Nt, (**e**) Nd and (**f**) k.

In order to look the significance of solutal concentration profile s for combined parameter Ha, vortex viscosity parameter K, Dufour Lewis number Ld and regular Lewis number Le for two different values of k i.e., $k = 0.0$ and $k = 0.5$, Figure 3a–d are suggested. Figure 3a reveals that solutal concentration profile s gets maximum values for Ha for both $k = 0.0$ and $k = 0.5$. Since the combined parameter is a combination of both Hartmann number and porosity parameter. Since Hartmann number has primary relation with Lorentz force which enables the considered profile to enhance. However, the solutal concentration layer is relatively thicker in absence of viscoelastic parameter $k = 0.0$. The influence of K on s has been nominated in Figure 3b. Here a decreasing profile of s has been noted as K varies. Figure 3c reveals that solutal concentration profile s reached at peak as Ld increase. From Figure 3d, it is noted that s increases by increasing Le. Such a trend can be justified physically as the Lewis factor reversely reflects with mass diffusivity. The diffusion rate becomes poor for enlarged values of Le which leads to the decrement of the resulting solutal concentration.

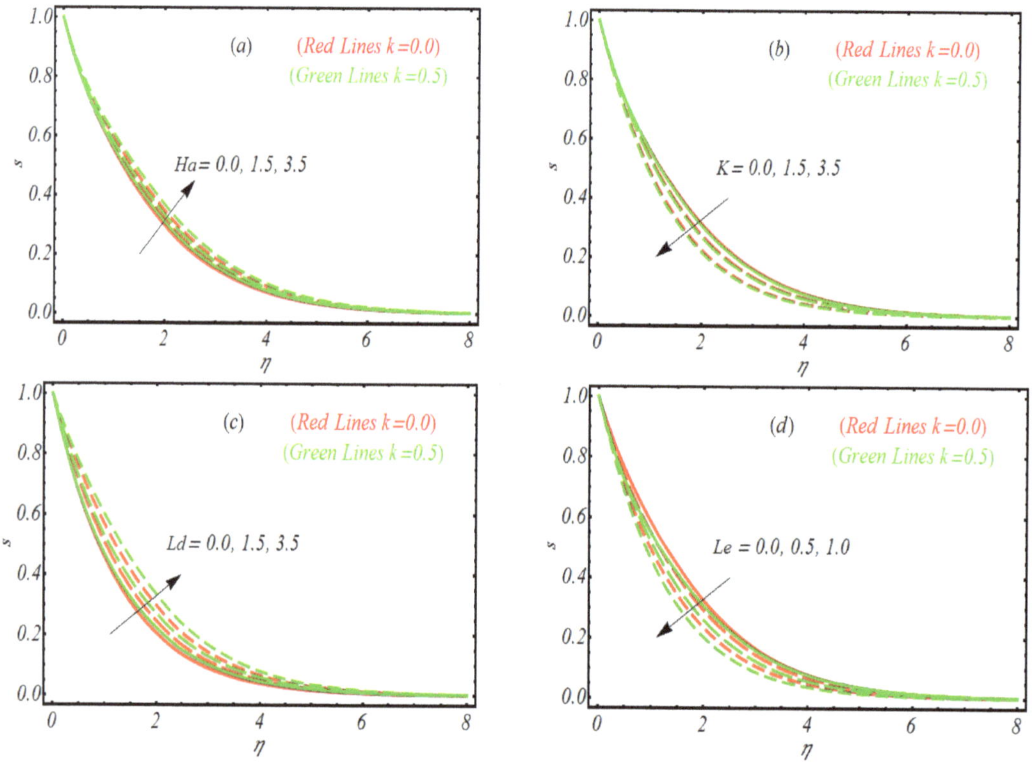

Figure 3. Solutal concentration profile for (**a**) Ha, (**b**) K, (**c**) Ld and (**d**) Le.

To examine how various involved parameters like vortex viscosity parameter K, nano Lewis number Ln, Brownian motion parameter Nb and thermophoresis parameter Nt influences the concentration of nanoparticles ϕ, Figure 4a–d is prepared. First, we look the graphical illustration of K on ϕ. Like the previous profile, the variation in ϕ for various parameters is observed for the highly viscoelastic case $k = 0.5$ and in absence of second grade fluid, $k = 0.0$. Figure 4a reveals that with the increase of K, the concentration profile for both $k = 0.0$ and $k = 0.5$ decrease effectively. Physically higher values of K are concerned with relatively poor viscosity and hence ϕ declined. Similar to the temperature and solutal concentration profile, the concentration profile is also relatively largely reduced for $k = 0.5$. Figure 4b punctuates that ϕ declines as Ln gets varied. The explanation for the profile of ϕ for Nb is notified in Figure 4c. Again, the larger values associated with Nb depressed the concentration of nanoparticles. The dependence of Nt on ϕ is discussed via Figure 4d. It is easily noted that concentration profile, by varying Nt, enhances. However, for $k = 0.0$, the concentration boundary layer is relatively thin as compared to $k = 0.5$. The results of Figure 5a,b pronounces the change of the microprism profile χ for Peclet number Pe and bioconvection Lewis number Lb. Both parameters reduce the microorganism profile.

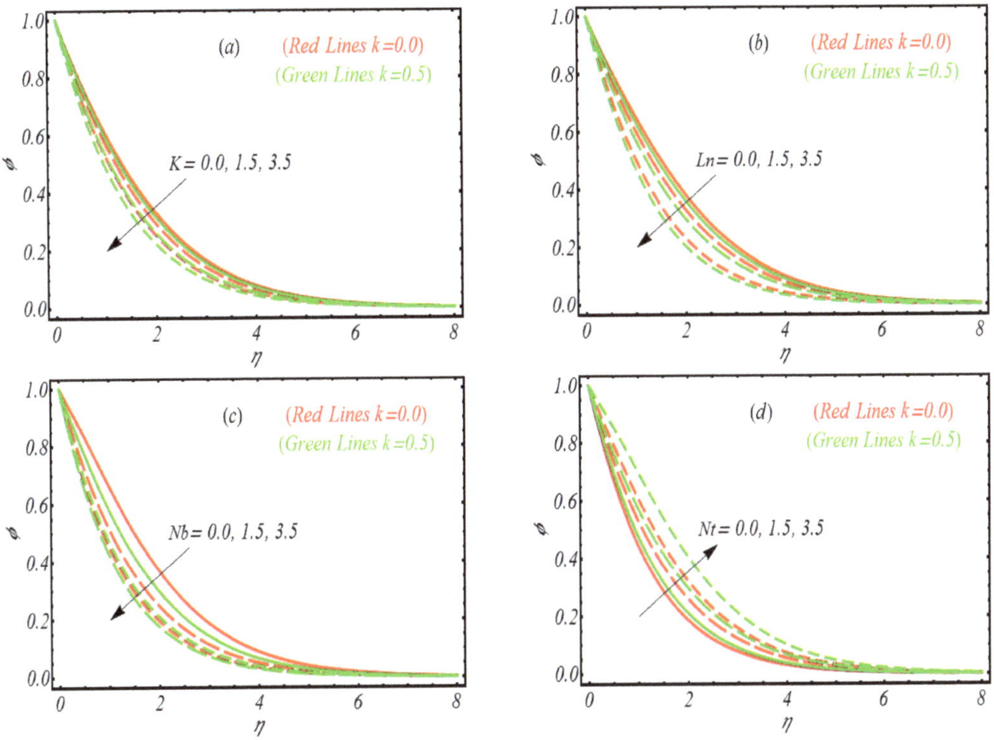

Figure 4. Concentration profile for (**a**) K, (**b**) Ln, (**c**) Nb and (**d**) Nt.

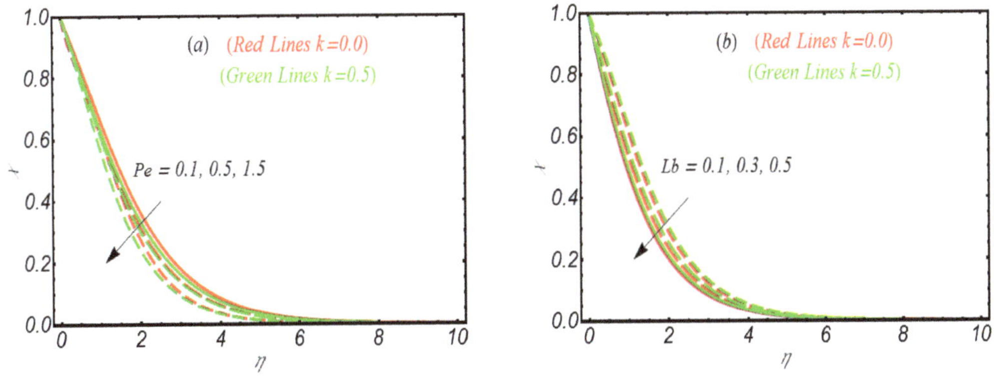

Figure 5. Microorganism profile for (**a**) Pe, (**b**) Lb.

The numerical data for of $f''(0)$ and $-g'(0)$ for combined parameter Ha and viscoelastic parameter k are reported in Table 3. Here it is observed that both $f''(0)$ and $-g'(0)$ increases for both Ha and k. However, the such demising variation in $-g'(0)$ is slower as compared to $f''(0)$. Table 4 is made to understand the influence of involved parameters like Ha, Pr, Nt, Nb, ε and K on Nusselt physical quantity, Sherwood constant and motile density number when $k = 0.1$. Here an increasing trend for these quantities has been noted for Pr and K while low numerical values are reported for Ha, Nt, Nb and ε. Studies [39–41] highlight the physical significance of heat transfer rate and flow mechanisms [42,43] over diverse surfaces.

Table 3. Numerical values of $f''(0)$ and $-g'(0)$ for number Ha and k.

Ha	k	$f'(0)$	$-g'(0)$
0.1		0.763453	0.0542344
0.3	0.2	0.810542	0.0585465
0.7		0.845234	0.0621414
	0.3	0.78522	0.042234
0.2	0.5	0.823455	0.051355
	0.7	0.893443	0.5579554

Table 4. Numerical values of local Nusselt number and local Sherwood number when $k = 0.5$.

Ha	Pr	Nt	Nb	ε	K	$-\theta'(0)$	$-\phi'(0)$	$-\chi'(0)$
0.0						0.67456	0.52678	0.5056578
0.5	0.7	0.3	0.3	0.1	0.1	0.63345	0.48635	0.46658
1.0						0.60945	0.44743	0.43665
	0.2					0.433435	0.464367	0.46065
0.5	0.7					0.48467	0.51046	0.48456
	1.0					0.510535	0.54426	0.49547
		0.0				0.432345	0.550768	0.452075
	0.7	0.3				0.4165	0.50235	0.43867
		0.5				0.3978	0.46436	0.40655
			0.1			0.49361	0.54646	0.4568
		0.3	0.3			0.45546	0.58643	0.49846
			0.6			0.43436	0.60266	0.51754
				0.0		0.440241	0.492345	0.4345456
				0.3		0.43656	0.46654	0.4054545
				0.6		0.40567	0.43742	0.391123
					0.0	0.4437	0.50436	0.50423
				0.1	0.4	0.5159	0.53576	0.53366
					0.7	0.53755	0.56747	0.570547

7. Conclusions

This continuation directed an analytical approach to the double diffusion flow of viscoelastic micropolar fluid by utilizing nanoparticles. Thermophoresis and Brownian effects are also elaborated on in the current analysis. Physical phenomena are formulated by using boundary layer approximations and, later on, analytical simulations have been suggested by using HAM. Some further studies on fluid flow are listed in [44–48]. Main results are given as

a. The thermal profile decreases by increasing vortex viscosity parameters while the opposite behavior is visualized for the variable thermal conductivity parameter and thermophoresis parameter. Such a variation trend is more dominant for viscoelastic cases.

b. The combined parameter and Dufour Lewis number progressively enlarge the heating and solutal concentration phenomena.
c. The lower effects for volume fraction of nanofluid due to nano Lewis number are noted.
d. The Nusselt quantity and Sherwood number declined for variable thermal conductivity parameters and thermophoresis parameters.
e. The observations made here may play vital role for suggesting and improving manufacturing and diffusion processes.

Author Contributions: Conceptualization, A.A. and S.U.K.; methodology, C.G.; software, K.A.-K.; validation, M.I.K., O.T.B. and K.G.; formal analysis, S.M.E.; investigation, A.A.; resources, M.I.K.; data curation, S.U.K. and O.T.B.; writing—original draft preparation, A.A., S.U.K. and M.I.K.; writing—review and editing, K.G. and K.A.-K.; visualization, O.T.B. and C.G.; supervision, M.I.K. and S.U.K.; project administration, M.I.K.; funding acquisition, S.M.E. and M.I.K. All authors have read and agreed to the published version of the manuscript.

Funding: The authors would like to thank the Deanship of Scientific Research at Umm Al-Qura University for supporting this work by Grant Code: 23UQU4331317DSR110.

Institutional Review Board Statement: Not applicable.

Informed Consent Statement: Not applicable.

Data Availability Statement: All the data are available inside the manuscript.

Conflicts of Interest: The authors declare no conflict of interest.

Nomenclature

u	velocity
b	stretching constant constant
H_a	combined parameter
k	viscoelastic parameter
Nt	thermophoresis parameter
Nd	modified Dufour number
Pe	Peclet number
δ_1	microorganism concentration difference
Ln	nano Lewis number
ε	thermal dependence conductivity factor
σ_e	electrical conductivity
ϑ	stands for permeability of porous medium
ρ_f	fluid density
N	micro-rotation
D_B	Brownian diffusion coefficients
D_s	solutal diffusivity
DK_{CT}	Soret diffusivity
C	concentration
T	temperature
K	vortex viscosity parameter
Pr	Prandtl number
Nb	Brownian motion parameter
Le	regular Lewis number
Ld	Dufour Lewis number
Lb	bioconvection Lewis
k_1	fluid constant
K_∞	material conductivity
k_0	viscoelastic parameter
γ	spin gradient viscosity
j	microinertia per unit mass
α_1	thermal diffusivity

DK_{TC}	Dufour diffusivity
D_T	thermophoretic diffusion coefficient
$\tau_T = (\rho c)_p/(\rho c)_f$	heat capacity of nanoparticles to heat capacity of fluid ratio.

References

1. Choi, S.U.S. Enhancing thermal conductivity of fluids with nanoparticles. *Int. Mech. Eng. Cong. Exp.* **1995**, 99–105.
2. Buongiorno, J. Convective Transport in Nanofluids. *J. Heat Transfer.* **2006**, *128*, 240–250. [CrossRef]
3. Ramana Reddy, J.V.; Sugunamma, V.; Sandeep, N. Thermophoresis and Brownian motion effects on unsteady MHD nanofluid flow over a slendering stretching surface with slip effects. *Alex. Eng. J.* **2018**, *57*, 2465–2473. [CrossRef]
4. Usman, M.; Soomro, F.A.; Haq, R.U.; Wang, W.; Defterli, O. Thermal and velocity slip effects on Casson nanofluid flow over an inclined permeable stretching cylinder via collocation method. *Int. J. Heat Mass Transf.* **2018**, *122*, 1255–1263. [CrossRef]
5. Yin, C.; Zheng, L.; Zhang, C.; Zhang, X. Flow and heat transfer of nanofluids over a rotating disk with uniform stretching rate in the radial direction. *Propuls. Power Res.* **2017**, *6*, 25–30. [CrossRef]
6. Madhu, M.; Kishan, N.; Chamkha, A.J. Unsteady flow of a Maxwell nanofluid over a stretching surface in the presence of magnetohydrodynamic and thermal radiation effects. *Propuls. Power Res.* **2017**, *6*, 31–40. [CrossRef]
7. Turkyilmazoglu, M. Free and circular jets cooled by single phase nanofluids. *Eur. J. Mech.-B/Fluids* **2019**, *76*, 1–6. [CrossRef]
8. Turkyilmazoglu, M. Buongiorno model in a nanofluid filled asymmetric channel fulfilling zero net particle flux at the walls. *Int. J. Heat Mass Transf.* **2018**, *126*, 974–979. [CrossRef]
9. Abbasi, F.M.; Shanakhat, I.; Shehzad, S.A. Entropy generation analysis for peristalsis of nanofluid with temperature dependent viscosity and Hall effects. *J. Magn. Magn. Mater.* **2019**, *474*, 434–441. [CrossRef]
10. Waqas, H.; Shehzad, S.A.; Khan, S.U.; Imran, M. Novel Numerical Computations on Flow of Nanoparticles in Porous Rotating Disk with Multiple Slip Effects and Microorganisms. *J. Nanofluids* **2019**, *8*, 1423–1432. [CrossRef]
11. Sun, L.; Wang, G.; Zhang, C.; Jin, Q.; Song, Y. On the rheological properties of multi-walled carbon nano-polyvinylpyrrolidone/silicon-based shear thickening fluid. *Nanotechnol. Rev.* **2021**, *10*, 1339–1348. [CrossRef]
12. Li, J.; Liang, Z.; Chen, Z.; Zhang, Z.; Liu, H.; Liu, Z.; Xu, Z. Engineering unsaturated sulfur site in three-dimension MoS2@rGO nanohybrids with expanded interlayer spacing and disordered structure for gaseous elemental mercury trap. *Chem. Eng. J.* **2023**, *453*, 139767. [CrossRef]
13. Li, Z.; Zhang, Q.; Shen, H.; Xiao, X.; Kuai, H.; Zheng, J. Buckling performance of the encased functionally graded porous composite liner with polyhedral shapes reinforced by graphene platelets under external pressure. *Thin-Walled Struct.* **2023**, *183*, 110370. [CrossRef]
14. Yang, Y.; Wang, S.; Wen, H.; Ye, T.; Chen, J.; Li, C.; Du, M. Nanoporous Gold Embedded ZIF Composite for Enhanced Electrochemical Nitrogen Fixation. *Angew. Chem. Int. Ed.* **2019**, *58*, 15362–15366. [CrossRef]
15. Agarwal, S.; Rana, P. Analysis of periodic and aperiodic convective stability of double diffusive nanofluid convection in rotating porous layer. *Appl. Math. Mech.* **2016**, *37*, 215–226. [CrossRef]
16. Raghunatha, K.R.; Shivakumara, I.S. Stability of triple diffusive convection in a viscoelastic fluid-saturated porous layer. *Appl. Math. Mech.* **2018**, *39*, 1385–1410. [CrossRef]
17. Daba, M.; Devaraj, P. Unsteady double diffusive mixed convection flow over a vertically stretching sheet in the presence of suction/injection. *J. Appl. Mech. Tech. Phys.* **2017**, *58*, 232–243. [CrossRef]
18. Gireesha, B.; Archana, M.; Prasannakumara, B.C.; Gorla, R.R.; Makinde, O.D. MHD three dimensional double diffusive flow of Casson nanofluid with buoyancy forces and nonlinear thermal radiation over a stretching surface. *Int. J. Numer. Methods Heat Fluid Flow* **2017**, *27*, 2858–2878. [CrossRef]
19. Bilal, S.; Rehman, K.U.; Malik, M.; Hussain, A.; Awais, M. Effect logs of double diffusion on MHD Prandtl nano fluid adjacent to stretching surface by way of numerical approach. *Results Phys.* **2017**, *7*, 470–479. [CrossRef]
20. Hayat, T.; Qayyum, S.; Shehzad, S.A.; Alsaedi, A. Cattaneo–Christov double-diffusion model for flow of Jeffrey fluid. *J. Braz. Soc. Mech. Sci. Eng.* **2017**, *39*, 4965–4971. [CrossRef]
21. Irfan, M.; Khan, M.; Khan, W.A. On model for three-dimensional Carreau fluid flow with Cattaneo–Christov double diffusion and variable conductivity: A numerical approach. *J. Braz. Soc. Mech. Sci. Eng.* **2018**, *40*, 577. [CrossRef]
22. Prasannakumara, B.; Reddy, M.G.; Thammanna, G.; Gireesha, B. MHD Double-diffusive boundary-layer flow of a Maxwell nanofluid over a bidirectional stretching sheet with Soret and Dufour effects in the presence of radiation. *Nonlinear Eng.* **2018**, *7*, 195–205. [CrossRef]
23. Eringen, A.C. *Microcontinuum Field Theories*; Springer: New York, NY, USA, 2001; Volume I, II.
24. Eringen, A.C. Simple micro fluids. *Int. J. Eng. Sci.* **1964**, *2*, 205–217. [CrossRef]
25. Eringen, A.C. Theory of micropolar fluid. *J. Math. Mech.* **1966**, *16*, 1–18. [CrossRef]
26. Eringen, A.C. Theory of thermomicro fluids. *J. Math. Anal. Appl.* **1972**, *38*, 480–496. [CrossRef]
27. Nazeer, M.; Hussain, F.; Khan, M.I.; Khalid, K. Theoretical analysis of electrical double layer effects on the multiphase flow of Jeffrey fluid through a divergent channel with lubricated walls. *Waves Random Complex Media* **2022**, 1–15. [CrossRef]
28. Abbas, Z.; Sheikh, M.; Sajid, M. Hydromagnetic stagnation point flow of a micropolar viscoelastic fluid towards a stretching/shrinking sheet in the presence of heat generation. *Can. J. Phys.* **2014**, *92*, 1113–1123. [CrossRef]

29. Turkyilmazoglu, M. Mixed convection flow of magnetohydrodynamic micropolar fluid due to a porous heated/cooled deformable plate: Exact solutions. *Int. J. Heat Mass Transf.* **2017**, *106*, 127–134. [CrossRef]
30. Sui, J.; Zhao, P.; Cheng, Z.; Zheng, L.; Zhang, X. A novel investigation of a micropolar fluid characterized by nonlinear constitutive diffusion model in boundary layer flow and heat transfer. *Phys. Fluids* **2017**, *29*, 023105. [CrossRef]
31. Sajid, M.; Sadiq, M.; Ali, N.; Javed, T. Numerical simulation for Homann flow of a micropolar fluid on a spiraling disk. *Eur. J. Mech.-B/Fluids* **2018**, *72*, 320–327. [CrossRef]
32. Abdelmalek, Z.; Khan, S.U.; Awais, M.; Mustfa, M.S.; Tlili, I. Analysis of generalized micropolar nanofluid with swimming of microorganisms over an accelerated surface with activation energy. *J. Therm. Anal. Calorim.* **2020**, *144*, 1051–1063. [CrossRef]
33. Turkyilmazoglu, M. Determination of the correct range of physical parameters in the approximate analytical solutions of nonlinear equations using the Adomian decomposition method. *Mediterr. J. Math.* **2016**, *13*, 4019–4037. [CrossRef]
34. Liao, S.J. *Advance in the Homotopy Analysis Method. 5 Toh Tuck Link*; World Scientific Publishing: Singapore, 2014.
35. Khan, S.; Ali, N.; Hayat, T. Analytical and Numerical Study of Diffusion of Chemically Reactive Species in Eyring-Powell Fluid over an Oscillatory Stretching Surface. *Bulg. Chem. Commun.* **2017**, *49*, 320–330.
36. Khan, S.U.; Shehzad, S.A.; Ali, N. Interaction of magneto-nanoparticles in Williamson fluid flow over convective oscillatory moving surface. *J. Braz. Soc. Mech. Sci. Eng.* **2018**, *40*, 195. [CrossRef]
37. Turkyilmazoglu, M. The analytical solution of mixed convection heat transfer and fluid flow of a MHD viscoelastic fluid over a permeable stretching surface. *Int. J. Mech. Sci.* **2013**, *77*, 263–268. [CrossRef]
38. El-Kabeir, S. Hiemenz flow of a micropolar viscoelastic fluid in hydromagnetics. *Can. J. Phys.* **2005**, *83*, 1007–1017. [CrossRef]
39. Shi, M.; Zhu, H.; Chen, C.; Jiang, J.; Zhao, L.; Yan, C. Synergistically coupling of graphene quantum dots with Zn-intercalated MnO2 cathode for high-performance aqueous Zn-ion batteries. *Int. J. Miner. Met. Mater.* **2022**, *30*, 25–32. [CrossRef]
40. Feng, X.; Jiang, L.; Li, D.; Tian, S.; Zhu, X.; Wang, H.; He, C.; Li, K. Progress and key challenges in catalytic combustion of lean methane. *J. Energy Chem.* **2022**, *75*, 173–215. [CrossRef]
41. Huang, Z.; Li, T.; Huang, K.; Ke, H.; Lin, M.; Wang, Q. Predictions of flow and temperature fields in a T-junction based on dynamic mode decomposition and deep learning. *Energy* **2022**, *261*, 125228. [CrossRef]
42. Yang, J.; Fu, L.-Y.; Zhang, Y.; Han, T. Temperature- and Pressure-Dependent Pore Microstructures Using Static and Dynamic Moduli and Their Correlation. *Rock Mech. Rock Eng.* **2022**, *55*, 4073–4092. [CrossRef]
43. Xiao, D.; Hu, Y.; Wang, Y.; Deng, H.; Zhang, J.; Tang, B.; Xi, J.; Tang, S.; Li, G. Wellbore cooling and heat energy utilization method for deep shale gas horizontal well drilling. *Appl. Therm. Eng.* **2022**, *213*, 118684. [CrossRef]
44. Chu, Y.-M.; Khan, M.I.; Abbas, T.; Sidi, M.O.; Alharbi, K.A.M.; Alqsair, U.F.; Khan, S.U.; Malik, M. Radiative thermal analysis for four types of hybrid nanoparticles subject to non-uniform heat source: Keller box numerical approach. *Case Stud. Therm. Eng.* **2022**, *40*, 102474. [CrossRef]
45. Bafakeeh, O.T.; Raza, A.; Khan, S.U.; Khan, M.I.; Nasr, A.; Ben Khedher, N.; Tag-Eldin, E.S.M. Physical Interpretation of Nanofluid (Copper Oxide and Silver) with Slip and Mixed Convection Effects: Applications of Fractional Derivatives. *Appl. Sci.* **2022**, *12*, 10860. [CrossRef]
46. Ali, Q.; Al-Khaled, K.; Omar, J.; Raza, A.; Khan, S.U.; Khan, M.I.; Najati, S.A.; Oreijah, M.; Guedri, K.; Galal, A.M. Analysis for advection–diffusion problem subject to memory effects and local and nonlocal kernels: A fractional operators approach. *Int. J. Mod. Phys. B* **2022**, 2350099. [CrossRef]
47. Ali, Q.; Al-Khaled, K.; Khan, M.I.; Khan, S.U.; Raza, A.; Oreijah, M.; Guedri, K. Diffusion phenomenon for natural convection flow of classical Hartmann problem due to a cylindrical tube by generalized Fourier's theories: A Fractional analysis. *Int. J. Mod. Phys. B* **2022**, 2350104. [CrossRef]
48. Harshavardhan, A.; Matt, S.B.; Khan, M.I.; Prakash, K.K.; Alnuwaiser, M.A.; Betageri, V.S.; Sidlinganahalli, M. Synthesis of Tin oxide nanoparticles using Nelumbo nucifera leaves extract for electrochemical sensing of dopamine. *Int. J. Mod. Phys. B* **2022**. [CrossRef]

bioengineering

Article

Holistic Design of Experiments Using an Integrated Process Model

Thomas Oberleitner, Thomas Zahel, Barbara Pretzner and Christoph Herwig

[1] Competence Center CHASE GmbH, Ghegastraße 3, Top 3.2, 1030 Vienna, Austria
[2] Körber Pharma Austria GmbH, PAS-X Savvy, Mariahilferstraße 88A/1/9, 1070 Vienna, Austria
[3] Research Area Biochemical Engineering, Vienna University of Technology, Gumpendorferstrasse 1a, 1060 Vienna, Austria
* Correspondence: christoph.herwig@tuwien.ac.at

Abstract: Statistical experimental designs such as factorial, optimal, or definitive screening designs represent the state of the art in biopharmaceutical process characterization. However, such methods alone do not leverage the fact that processes operate as a mutual interplay of multiple steps. Instead, they aim to investigate only one process step at a time. Here, we want to develop a new experimental design method that seeks to gain information about final product quality, placing the right type of run at the right unit operation. This is done by minimizing the simulated out-of-specification rate of an integrated process model comprised of a chain of regression models that map process parameters to critical quality attributes for each unit operation. Unit operation models are connected by passing their response to the next unit operation model as a load parameter, as is done in real-world manufacturing processes. The proposed holistic DoE (hDoE) method is benchmarked against standard process characterization approaches in a set of in silico simulation studies where data are generated by different ground truth processes to illustrate the validity over a range of scenarios. Results show that the hDoE approach leads to a >50% decrease in experiments, even for simple cases, and, at the same time, achieves the main goal of process development, validation, and manufacturing to consistently deliver product quality.

Keywords: design of experiments; holistic experimental design; integrated process model; optimal designs; process characterization; biopharmaceutical process validation

Citation: Thomas Oberleitner, Thomas Zahel, Barbara Pretzner and Christoph Herwig Holistic Design of Experiments Using an Integrated Process Model. *Bioengineering* **2022**, *9*, 643. https://doi.org/10.3390/bioengineering9110643

Academic Editors: Sundeep Singh, Roderick Melnik and Esther Pueyo

Received: 14 October 2022
Accepted: 1 November 2022
Published: 3 November 2022

1. Introduction

The goal of process characterization in biopharmaceutical development is to establish scientific evidence that a process is able to consistently deliver quality products. An important part of this procedure is to determine the effect of process parameters (PP) on critical quality attributes (CQA [1,2]). Design of experiments (DoE) is a well-established tool to design experimental runs that yield such information and is oftentimes followed by data analysis and inference based on regression models [3,4]. DoE variants such as factorial or optimal designs facilitate the detection of effects by minimizing or eliminating correlation, and they are comprised of all possible combinations of effect levels or a subset thereof [5]. These experiments are then conducted in small-scale models for each unit operation (UO), and results can be used to create mathematical models that quantify the impact of effects.

Of particular interest when defining a control strategy is the range in which PPs can safely operate while keeping CQA concentrations within acceptable boundaries. Those proven acceptable ranges (PAR) are part of the control strategy a manufacturer might submit to a regulatory authority [6]. One way of defining a parameter's PAR is illustrated in Figure 1. Using this approach, the PAR is affected by two components: first, the model prediction, including the statistical intervals and second, the acceptance limits, as the intersection between the two defines the PAR's range. Another option could be to alter the setpoint condition of other PPs (see Section 4.2). This leads to a change in the univariate prediction plot by shifting the prediction vertically (orange line in Figure 1). However, optimization by changing setpoint conditions is usually not the focus of process characterization. The PAR is required to allow for sufficient process and operator variability while being conservative enough to keep CQAs within acceptance limits. If the PAR is too narrow for adequate operability, one can either try to reduce model uncertainty by investing DoE runs or change acceptance limits by performing spiking runs.

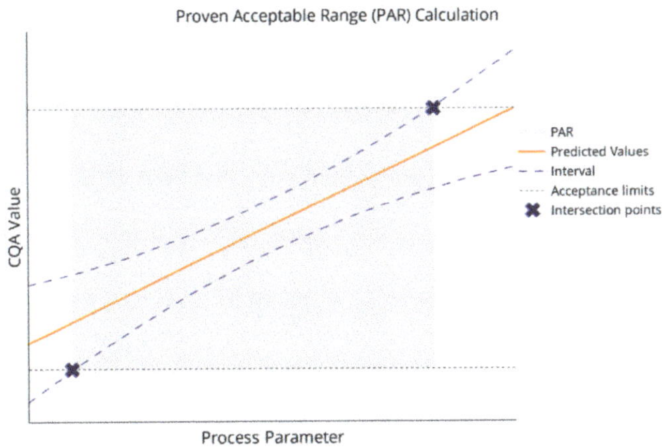

Figure 1. An example for how the PAR of a process parameter can be calculated. The predicted mean of the CQA as a function of the PP is shown in orange and the statistical interval around these predictions is illustrated as dashed lines. Lower and upper PAR boundaries can be defined by the intersection points of a statistical interval and the CQA acceptance criteria, marked as x.

1.1. *Option A: Improving Model Estimates via DoE*

To illustrate the effect of DoE runs on model uncertainty, consider the formula for calculating tolerance intervals for a normally distributed population [7]:

$$\hat{y} \pm z_{\frac{1+\psi}{2}} \sqrt{\frac{\nu\left(1+\left(\frac{1}{N}\right)\right)}{\chi^2_{\alpha,\nu}}}\sigma \tag{1}$$

$\hat{y}$ is the mean prediction of the model, $1-\psi$ the nominal proportion of the population covered by the interval, and α the confidence level. Disregarding the critical value for the normal distribution $z_{\frac{1+\psi}{2}}$ and the standard deviation σ, the dominant factor in this formula is the square root term, which includes the lower α quantile of the χ^2 distribution in its denominator and converges toward one as N increases. The residual degrees of freedom in a regression model are calculated as $\nu = N - p$, where p is the number of model parameters. Note that this is a simplified version of the tolerance interval, and other methods might be used in a regression setting [8–10]. The graph of the square root term is shown in Figure 2 over a range of possible values of N.

The figure illustrates the strong decrease in this factor for the first values of N before the curve starts to flatten. While other measures of model quality, e.g., the standard deviation or parameter covariance are contributing factors in a regression setting, this effect is representative of the behavior of an interval as the number of observations increases. For the experimental effort invested in improving parameter estimates and model quality this means that at some point no large improvements can be achieved in the interval width and, in turn, the PAR. Then, tackling the second decisive element, the intermediate acceptance criteria, might be more rewarding.

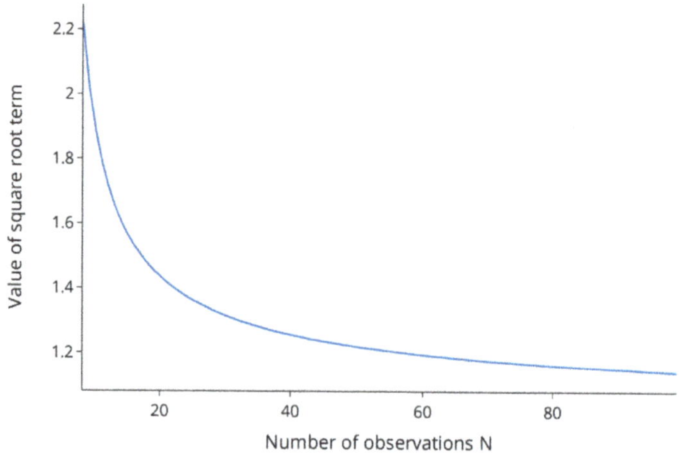

Figure 2. Value of the term $\sqrt{\frac{v(1+(\frac{1}{N}))}{\chi^2_{\alpha,v}}}$ over a range of values for N.

1.2. Option B: Improving Acceptance Limits via Spiking Studies

In typical biopharmaceutical process development and characterization, unit operations are studied individually, and acceptance criteria need to be defined for each UO. Those intermediate acceptance criteria (iAC) are the second component affecting the PAR calculation, as shown in Figure 1. A frequently followed but flawed approach to setting iACs is to calculate this range using three standard deviations (SD) of manufacturing scale runs [11–13]. A much more scientifically sound method has recently been published where the only requirement is to have drug substance/product specification. In this approach, an IPM is used, and iACs for all UOs can be calculated inversely, starting from the specifications [11]. The same approach is shown here in an illustrative manner: one can correlate the inputs/loads and outputs/pools of each unit operation, as shown in Figure 3. If the slope of this correlation equals one, all of the load will be found in the pool, which is not desirable in a downstream UO. If the slope equals zero, the same (low) pool value will be achieved regardless of the load values, which is an ideal and robust scenario of a downstream UO. We can now calculate the iAC by backpropagating the iAC of the next UO (starting with the DS specifications) through those models. By iteratively applying this technique, the iACs for the entire process can be calculated. As all models are data-based, conservative extrapolation needs to be taken into account when making predictions outside the observed training space of the explanatory variables (here, the input material of each model). As discussed in [14], for impurities, we assume that every additional amount of loaded impurity beyond the observed range will be directly propagated into the output/pool of the UO, which can be mathematically seen as a piecewise regression model with a slope of one (see Figure 3).

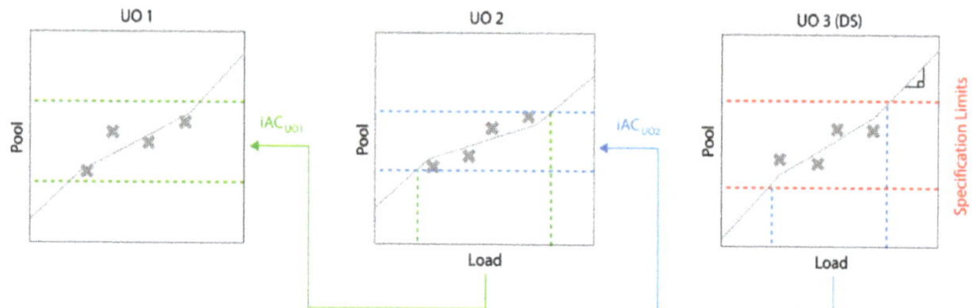

Figure 3. Inverse iAC calculation, beginning with drug specifications at the last UO.

Figure 4 illustrates the effect of adding a spiking run that shows successful clearance. The data point is added at UO 3 and the observed trend will be extrapolated, leading to an increase in iAC in UO 2. Since the iAC of UO 2 is used to calculate the iAC of UO 1, its iAC will be increased as well, etc. Hence, introducing a spiking run at one UO will potentially lead to increase in iACs of all previous UOs. Of course, the addition of spiking runs at extreme levels will be limited by the clearance capacity of the downstream UO. Note that Figures 3 and 4 show a simplified version for calculating iACs and that the actual method can involve more advanced statistical methods such as tolerance intervals and Monte Carlo sampling (for details, see [11]). Furthermore, while a linear correlation is assumed here, any mathematical model can be used to describe the dependency between input and output of individual UOs.

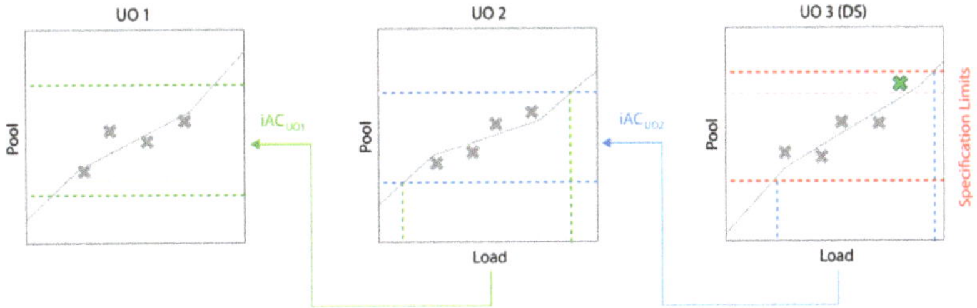

Figure 4. The effect of adding an additional data point on the iAC calculation.

Using spiking runs is not a novelty and commonly used in the industry [15,16]. However, results are usually only reported in documents and in our experience do not find entrance into mathematical modeling that also accounts for the uncertainty around individual experiments. Marschall et al. describes how any manufacturing or small-scale data can be used to calculate iACs and how spiking runs are included in that procedure [11].

At this point, we have demonstrated that both the addition of DoE runs to decrease model uncertainty as well as the addition of spiking runs to increase iACs can help to gain process understanding, which helps to increase PARs and facilitates a more flexible control strategy. However, it remains unclear which combination of DoE or spiking runs would give the maximum gain in PAR. Therefore, we want to:

- Develop a recommender system, called the holistic design of experiments (hDoE), that suggests the optimal runs (DoE or spiking) at specific UOs that lead to the fastest increase in process understanding. Here we define process understanding as the accuracy and precision of the (unknown) true relation between all PPs and CQAs, as well as the input/output relation of individual UOs. We describe this method in Section 3.1;
- Demonstrate that using such a recommender system can lead to a significant reduction in the required number of total runs of a process characterization study (PCS) using state-of-the-art workflows. We verify this in a set of simulation studies presented in Section 3.2;
- Identify an accelerated workflow for PCS using hDoE that can be applied in practice; see Section 4.1.

2. Materials and Methods

2.1. Optimal Designs

Factorial or fractional factorial designs are generally considered the ideal approach to creating experimental designs that yield the most information about how process parameters affect the response. An exhaustive account of such designs can be found in [5]. In practice, however, they are not universally applicable as they often require many runs and cannot incorporate existing data. Optimal designs constitute a more flexible alternative [17–20]. The number of runs required is not a consequence of the chosen type of design, e.g., 2^k, for a factorial design with two levels and k parameters but can be chosen more flexibly. Furthermore, optimal designs can be used to augment an existing set of runs. Both of those properties are important for their application in hDoE, as the procedure starts off with a minimal set of experiments far smaller than a full factorial design, which gets augmented in each experiment/evaluation cycle. Based on the working set of already performed experiments, new runs based on optimal designs are recommended. Optimal designs optimize specific properties of the design matrix X. For example, D-optimal designs minimize the variance of parameter estimates in a model. As $Var(\hat{\beta}) = \sigma^2 (X^T X)^{-1}$, this is equivalent to maximizing $|X^T X|$, the determinant of the squared design matrix and one can see that this is maximized when the columns of X are orthogonal. However, in contrast to factorial designs, strict orthogonality is not required and one consequence of that is that the number of runs in X can be set freely depending on the use case. The rows in X are then chosen by exchange algorithms from a candidate set of runs, which generally consists of all possible level combinations for the main effects defined in the model. We used augmented D-optimal designs to generate the experiments recommended by hDoE.

2.2. Integrated Process Model

An integrated or holistic process model is an in-silico representation of a manufacturing process comprised of multiple steps or unit operations. While there are many approaches to constructing process models (see [21] for an overview), here we consider the IPM as an empirical ensemble model implemented as a sequence of UO models that enable predictions of different CQAs as a function of process parameters. To simulate CQA concentrations, the predicted values are passed on to the next UO as a process parameter in a Monte Carlo simulation that randomly draws parameter values [22]. This is done over the entire chain of UOs in the process, from upstream operations to the final drug substance. For a comprehensive description of the method, please consult [14].

An important aspect of this approach is extrapolation. Because parameter values are drawn randomly from their corresponding distribution, CQA predictions of one UO regularly exceed the range of values observed in the training data of the predecessor UO. This is a problem for the conservative prediction of CQA concentrations, as UOs are modeled as data-driven regression models. Such models are generally only valid within the range of the training data and extrapolation can lead to highly biased results [23].

2.3. Simulation Study

We investigated the performance of hDoE in a set of simulation studies illustrating different situations and processes. Results are compared to a state-of-the-art (SOTA) process characterization workflow that consists of conducting experiments based on optimal designs to investigate the impact of PPs on CQAs per UO. For the reference method, a D-optimal design was chosen with 6, 12, and 23 runs per UO, which leads to 24, 48, or 92 runs overall in 4 UOs.

Each process in this simulation study consists of a sequence of unit operations represented by ground truth equations that map PPs to a CQA. The first UO can be interpreted as the fermentation step, followed by three downstream UOs. The equations that describe these UOs satisfy IPM conditions by adhering to the heredity principle [24] and having linear load dependencies. We then try to find effect coefficients in the presence of added noise, employing both hDoE and the SOTA method that uses a predefined number of runs per UO, as described above. For hDoE, we start with a minimal design of 6 runs per UO (total 24 runs for 4 UOs) and add an additional 30 runs, chosen by the recommender system. Note that for the simulation study results reported here, only a single run was recommended per cycle, though results are similar for larger sets of run recommendations. To calculate OOS rates, an upper drug substance specification for the output/pool of UO 4 was set as three standard deviations above the mean of the ground truth process. Hence, when adding an infinite number of runs, a minimum OOS rate of 0.00135 ([1–0.9973]/2) can be achieved. As the OOS rate simulated by the IPM is based on drawing random values from a PP distribution, mean and variance must be specified. In this normalized setting, each PP's setpoint, i.e., the mean of the distribution, was chosen to be zero and the variance was set to be the same as that of the observed ground truth data that were used to derive specification limits. This was kept constant over all simulated hDoE steps to avoid a misleading optimization trajectory that improves OOS rates by simply reducing PP variances without increasing process knowledge. Each simulation scenario was repeated 100 times with different random seeds. A summary of simulation parameters is provided in Table 1.

Table 1. Parameters for the simulation study.

Parameter	Value
Number of UOs	4
Parameters per UO	5
CQA type	Impurity
hDoE start runs	6
Noise/std ratio for residual error	0.5 (0.9 in study D)
Number of runs recommended per cycle	1
Variable selection method	Bi-directional stepwise
p-value threshold for including effect	0.25
p-values threshold for excluding effect	0.05

The number of hDoE steps, repetitions of the simulation studies and the bi-directional stepwise variable selection method [25] for (re)fitting IPM models were chosen as a compromise between accuracy of results and simulation runtime.

For demonstration purposes but without loss of generality, we employ all simulation studies only for one CQA of the product. However, the methodology is not limited to the number of investigated CQAs. In practice, one would focus on the CQA, which shows the highest OOS rate.

2.3.1. Study A: Baseline

This simulation study represents a typical biopharmaceutical process with some quadratic and interaction effects and coefficients commonly found in characterization studies. In our experience, approximately 20–40% of all possible effects are practically significant in a model. A total of 5 factors lead to 20 effects (main, 2-factor interaction, and quadratic). In the ground truth, we have chosen 4–6 active effects, which equals 20–30% of all possible effects and is, therefore, within the expectation of a representative biopharmaceutical process. Moreover, we assume linear load dependencies and heredity between main and higher-order effects, which is also representative of the authors' experience. The ground truth equations for this scenario are as follows:

$$y_{UO1} = 8.0 + 0.7\,x_1 + 0.6\,x_2 + 0.5\,x_3 - 0.4\,x_4 + 0.9\,x_2^2 + 0.8\,x_3^2 \tag{2}$$

$$y_{UO2} = 3.4 + 0.5\,y_{UO1} + 0.5\,x_2 + 0.3\,x_3 + 0.5\,x_5 + 0.7\,x_2x_3 + 0.4\,x_2x_5 \tag{3}$$

$$y_{UO3} = 3.0 + 0.3\,y_{UO2} + 0.4\,x_1 - 0.3\,x_2 + 0.2\,x_4 - 0.2\,x_5 + 0.3\,x_1^2 - 0.7\,x_5^2 \tag{4}$$

$$y_{UO4} = 2.8 + 0.2\,y_{UO3} + 0.1\,x_1 + 0.2\,x_3 + 0.2\,x_5 + 0.3\,x_1x_3 \tag{5}$$

2.3.2. Study B: Load Effect Set to One

As hDoE leverages the UOs dependency on the load, we investigated its behavior when the load coefficient is set to one, and its values are passed directly to the output of UO 3, provided other PPs are at setpoint. This mimics the situation where a full propagation of the CQA through this UO is expected, and no clearance takes place. This is, of course, not the desired behavior of a downstream unit operation. In this setting, the load coefficient in UO 3 was set to one:

$$y_{UO3} = 3.0 + 1.0\, \mathbf{y_{UO2}} + 0.4\, x_1 - 0.3\, x_2 + 0.2\, x_4 - 0.2\, x_5 + 0.3\, x_1^2 - 0.7\, x_5^2 \tag{6}$$

2.3.3. Study C: All Load Effects Set to One

Here we set the load effects of all UOs to one. Every UOs output is directly propagated to the next UO, which means that there is no information about the load that could be detected by spiking runs. This a very untypical scenario as usually, we expect some clearance activity of the downstream UOs (UO 2–4). The purpose of this simulation is to show that, in the worst case, hDoE performs similarly to standard approaches using a predefined set of runs.

$$y_{UO2} = 3.4 + 1.0\, \mathbf{y_{UO1}} + 0.5\, x_2 + 0.3\, x_3 + 0.5\, x_5 + 0.7\, x_2 x_3 + 0.4\, x_2 x_5 \tag{7}$$

$$y_{UO3} = 3.0 + 1.0\, \mathbf{y_{UO2}} + 0.4\, x_1 - 0.3\, x_2 + 0.2\, x_4 - 0.2\, x_5 + 0.3\, x_1^2 - 0.7\, x_5^2 \tag{8}$$

$$y_{UO4} = 2.8 + 1.0\, \mathbf{y_{UO3}} + 0.1\, x_1 + 0.2\, x_3 + 0.2\, x_5 + 0.3\, x_1 x_3 \tag{9}$$

2.3.4. Study D: Disabled Probability-Ratio-Threshold

To highlight the importance of the probability-ratio-threshold (PRT) decision scheme, a method borrowed from the Metropolis–Hastings algorithm that encourages the detection of new effects (described in Section 3.1), we repeat the baseline study without PRT. Here, the decision logic simply recommends the type of run that leads to the largest reduction in OOS. As PRT is most effective in situations where no clear decision can be made due to high residual error, the error/standard deviation ratio in the ground truth was increased from 0.5 to 0.9.

3. Results

3.1. Holistic Design of Experiments

In this contribution, we propose a new tool for process development and characterization: holistic design of experiments (hDoE), an iterative approach to experimental design and evaluation that minimizes the number of runs invested while maximizing the overall process understanding, as defined in the introduction. As all UOs of the process contribute to the generation of DS material, we can boil down process understanding to how well the true distribution of DS product quality is known. To express this in a single measure that can be used for optimization, we chose the out-of-specification (OOS) rate based on known DS specifications.

As described in Section 2.2, the IPM connects UOs by passing the output of a UO to the next one as a load parameter. While the output might be affected by different parameters and interactions, the univariate relationship between load and output is assumed to be linear inside the range of observed load values in the training data. However, in the Monte Carlo simulation of a CQA distribution, simulated load values might exceed this observed range, in which case the CQA value is handled conservatively to avoid predictions that are too optimistic (see piecewise load model in Taylor et al. [14]). This means that the simulated CQA distribution in drug substances is highly dependent on the range of load parameter values in the training data. Small variations in the training data will lead to a broadened CQA distribution, accounting for the uncertainty due to extrapolation. However, in many cases, missing information about load parameters can be supplemented by performing spiking studies where PPs are kept at setpoint, and only the load is varied. This is in contrast to classic DoE studies, which examine specific combinations of factors and disregard the load.

Spiking and DoE runs are the two different types of the experiment proposed by hDoE in this contribution. Assuming model parameters do not change from one iteration to the next, DoE runs are expected to improve general process parameter estimates, while spiking studies improve estimates of the model's load coefficient and reduce extrapolation in the IPM simulation. However, the assumption of unchanging model parameters is regularly violated when variable selection on newly acquired data results in a new model. As a consequence, the simulated OOS probability is not guaranteed to be improved in every iteration.

hDoE starts out by roughly characterizing the process with a minimal D-optimal design per UO that facilitates fitting the initial regression models. Based on the information acquired in this first step, a set of runs is proposed. This set consists of DoE and spiking runs and includes the target UOs in which to perform them. The runs are chosen by how much they would reduce the OOS probability calculated by the IPM. After conducting the proposed experiments, the IPM data are supplemented with new information, and a variable selection step updates models where appropriate.

As OOS predictions and therefore run suggestions are based on models found in previous steps, the process is biased toward already detected effects. Additionally, the decision rule is susceptible to noise, especially in early steps, and might consider the value of adding spiking or DoE runs equivalent. Of course, this is generally not true, as only DoE experiments enable the detection of new or interaction effects. To mitigate this bias and encourage the detection of new effects, we employ a technique based on a decision scheme used in the Metropolis–Hastings algorithm [26]. Let X be the design matrix of the data already incorporated into the IPM and x_{DOE} and $x_{spiking}$ be new DoE and spiking samples, respectively, chosen from a set of sample candidates that result in the lowest OOS probability. We then calculate the ratio of those probabilities $\alpha = P(x_{DOE}|X)/P(x_{spiking}|X)$, draw a number from a uniform distribution, $u \in [0,\ 1]$, and only suggest a spiking run when $u \geq \alpha$. This means that DoE runs are always recommended if $P(x_{DOE}|X) \geq P(x_{spiking}|X)$. Spiking runs, however, are only suggested when the improvement in OOS probability considerably exceeds that of a DoE run. In the following, we term this the probability-ratio-threshold (PRT) approach. In our evaluation of the algorithm, PRT generally circumvents the problem of selecting spiking runs unnecessarily or overlooking effects (see Section 3.2). Figure 5 illustrates the individual steps and decision processes involved in hDoE.

3.2. *Simulation Results*

3.2.1. Out-of-Specification Rates

Figure 6 shows that in most scenarios, a high process understanding (quantified as a low OOS rate) can be achieved with a much smaller number of experiments compared to the SOTA method that uses a fixed number of DoE runs (in this case, 24, 48 and 92). This effect is most pronounced in study A, where the mean OOS rate drops to ~2.5% after only six additional runs recommended by hDoE (30 total), whereas 92 D-optimal runs calculated beforehand result in an OOS rate of ~7.5% due to the lack of exploration of load-to-pool dependencies. At first glance, this might appear as an unfair comparison, but workflows applied in the industry generally do not incorporate spiking runs in a mathematical framework to achieve a specific goal, e.g., establishing a control strategy. The effect of an UOs dependency on the load can also be seen in the results of study B, where variation around the OOS rate is larger due to setting the load coefficient to one in UO 3 in the ground truth. This means that no additional information about the load can be acquired in this UO, and the advantage of systematically recommending spiking studies at this UO is softened. Study C represents the worst case for hDoE, where the load has no effect in any unit operation, resulting in OOS rates close to that of the reference method. The larger variation shown in study D is due to increased noise in the ground truth. This, of course, affects hDoE as well as the reference method. However, note that variation is drastically increased when PRT is disabled (purple, dotted lines) compared to the recommended procedure that uses it when deciding on runs (blue, dotted lines).

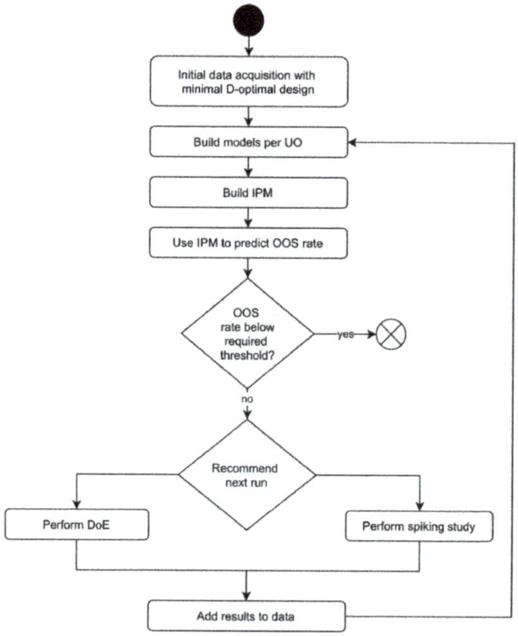

Figure 5. The hDoE recommender system that minimizes OOS by iteratively augmenting data and predicting optimal experiments.

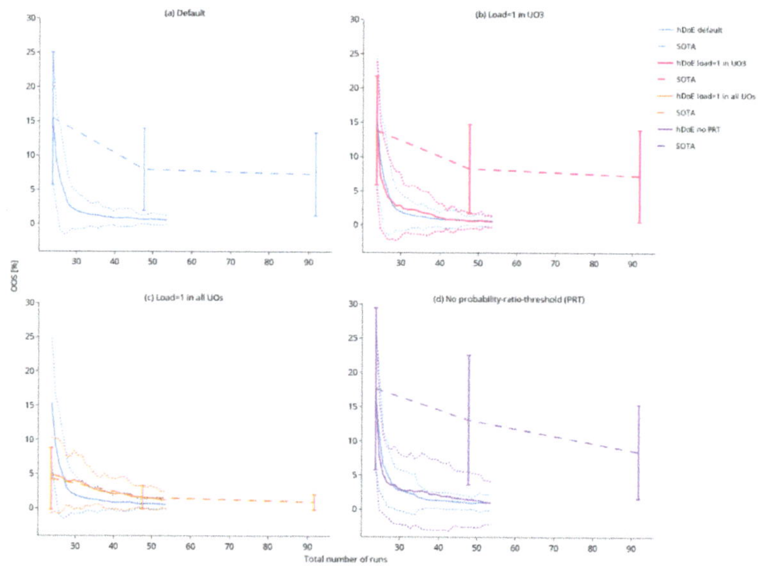

Figure 6. The decrease in OOS probability over the 30 hDoE steps for the different simulation scenarios. Each scenario was repeated 100 times. The solid line indicates the mean OOS probability, while the dotted lines show the standard deviation in the 100 repetitions of that step. For the reference method, OOS probabilities are plotted as dashed lines and were calculated at 24, 48, and 92 total runs.

3.2.2. Run Allocation

The drastic decrease in OOS probability over the number of experimental runs invested, shown in Figure 6, is achieved by performing spiking studies at the right UO. Generally, spiking runs are favored in the early steps of the procedure, as they eliminate extrapolation in the IPM (see Section 2.2) and therefore lead to the largest reduction in OOS early on. After this initial phase, larger OOS improvements can be achieved by improving parameter estimates, prompting hDoE to suggest more DoE runs. Figure 7 illustrates the allocation of different run types to the four unit operations of the simulation study. In the y-axis, the plots show the cumulative number of allocated runs over 100 repetitions of a simulation study, while the corresponding hDoE step can be seen in the x-axis. As the first UO (e.g., fermentation) is not affected by a load parameter, no spiking runs (dotted lines) are allocated. In the three simulation studies where the load influences a UO, spiking runs at UO 4 are recommended in the early steps and, in many of the 100 iterations, also in UO 3. This makes sense, as the OOS probability in drug substance, i.e., the last UO, is the main driver of the recommender system and its load coefficient directly affects the CQA distribution. In the absence of load effects, DoE runs are distributed approximately equally, while some spiking runs are accumulated due to noise (orange lines). The effect of increased noise and the absence of PRT are illustrated by the purple line. Much more spiking runs are invested, as the OOS simulation for the two types of experiment yields similar results, and choices are made more randomly.

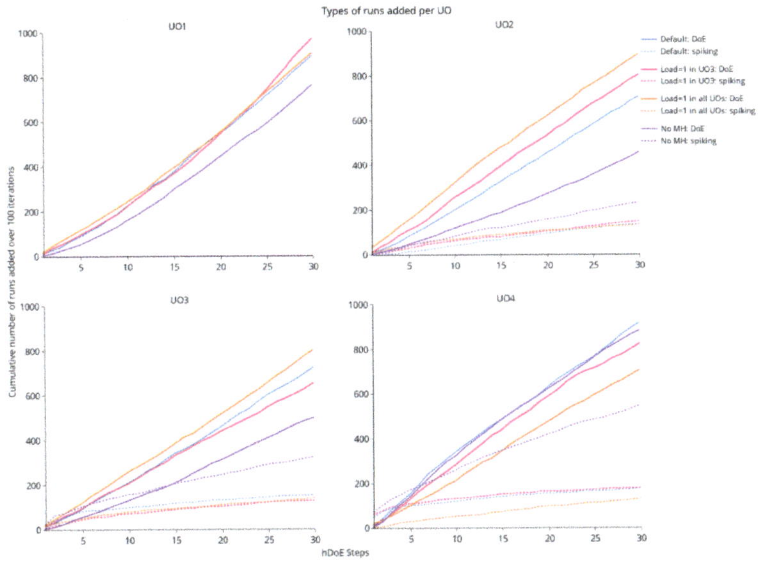

Figure 7. The cumulative allocation of either spiking or DoE runs in 100 repetitions of the simulation is shown over each step taken by the hDoE procedure (after the initial 24 runs). In total, each study distributes 100 × 30 = 3000 runs of any type to the four unit operations.

3.2.3. Parameter Estimates

A low OOS rate alone does not indicate correct models, as it does not account for aliasing effects in the parameters. Figure 8 shows the distributions of effects identified in the variable selection procedure at the last hDoE step (54 total runs invested). Each data point represents the effect size in a particular iteration of the simulation. We added an estimator of the residual variance in the form of the root mean squared error (RMSE). The mean estimated value (blue dashed line of each boxplot) converges to the ground truth (red solid line of each boxplot) for most parameters. Higher-order effect estimates in unit operations two and three are biased toward zero, i.e., they were not detected in variable selection. Note that stepwise variable selection was used for performance reasons in this simulation, which is known to eliminate effects prematurely [27], and that some of this bias could be mitigated by using more modern approaches such as leaps and bounds or other exhaustive algorithms. However, as tolerance intervals are used in the estimation of model uncertainty [14], overlooking individual effects, which results in larger estimates of the RMSE, is accounted for correctly in the uncertainty interval. Of course, the quality of parameter estimates is a direct consequence of the number of DoE runs invested, which in turn is dependent on when the hDoE procedure is terminated.

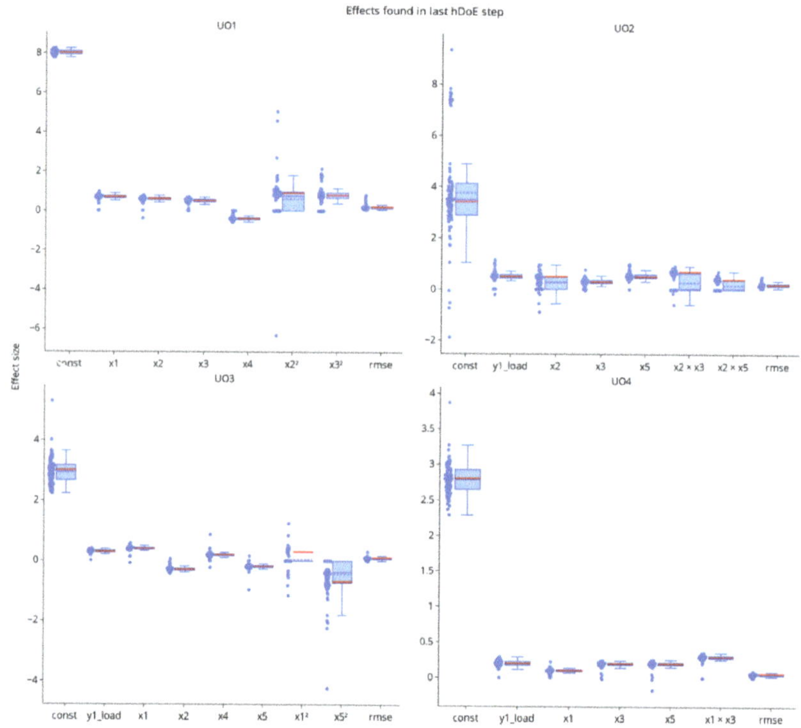

Figure 8. The distribution of effects found by stepwise variable selection at the last hDoE step. Each point represents the effect size in one of the 100 repetitions of the simulation study. The RMSE was added as an estimator of residual error.

Parameter estimates, especially for higher-order effects, are naturally not on par with those from a full D-optimal design with 23 runs per unit operation, see Figures 8 and 9. However, due to spiking studies invested in the characterization of the process, hDoE results in the improved estimation of load and intercept effects.

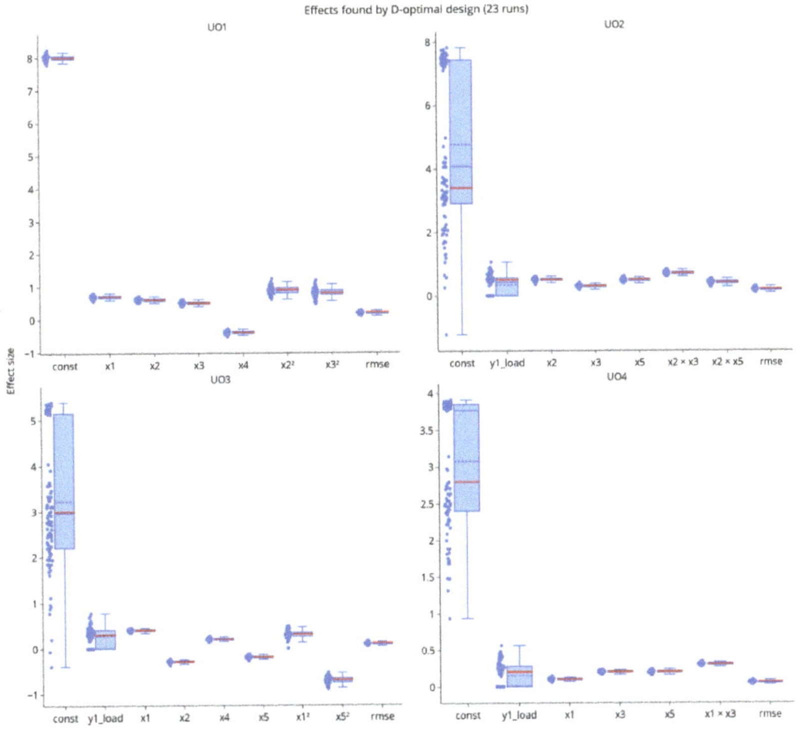

Figure 9. Parameter estimates for the full D-optimal design with 23 runs per UO. Variance in estimates is smaller than in hDoE. However, intercept and load effects are biased due to missing load data.

4. Discussion

4.1. hDoE in Process Characterization

Figure 10 shows how hDoE affects common steps in process characterization. The result of this procedure is a control strategy, of which PARs are an essential component. We describe in Section 1 the different courses of action when a PAR is too narrow to be part of an appropriate control strategy, steps that are also reflected by the hDoE recommender system. A third option that is currently not incorporated into the recommender system is to change the setpoint of other PPs that are active in the UO model. We present this approach as an outlook in Section 4.2. Finally, PAR ranges can be increased by reducing confidence/coverage levels of the statistical interval, although in most cases, this is not recommended and only mentioned here for the sake of completeness. We consider the workflow shown in Figure 10 as an extended version of the state-of-the-art workflow for process characterization (left column) that incorporates hDoE (right column, green boxes). This aligns with both the FDA and EMA guidelines for process characterization, as the former recommends DoE to increase process knowledge [3] and the latter impurity spiking challenges for downstream operations [28]. hDoE employed in the context of this extended workflow integrates both types of experiments and provides a systematic method for when and where to perform them.

4.2. Outlook: Changing PP Setpoints to Increase the PAR

An important aspect of the PAR calculation illustrated in Figure 4 is the univariate nature of this method. Only the screening range of the current PP is considered, while all other PPs are kept at their setpoint. As the UO model output, i.e., the CQA, is in most cases affected by multiple PPs, their setpoint can influence the offset of the univariate mean prediction of the target PP (orange line in the figure) significantly. Consequently, a change in the setpoint of another PP can push the predicted CQA distribution inside/outside the acceptance limits and change the OOS rate. Similarly, interaction effects with other PPs can also influence the target PPs effect.

The optimization of PP setpoints is already available in some statistical software [29], and hDoE could be easily extended to include such recommendations based on their effect on the OOS rate of the process. While this is generally not the focus of process characterization studies, the FDA recommends optimization based on setpoint shifts in the continuous verification phase [3]. This would advance the applicability of the hDoE recommender system into the domain of process optimization.

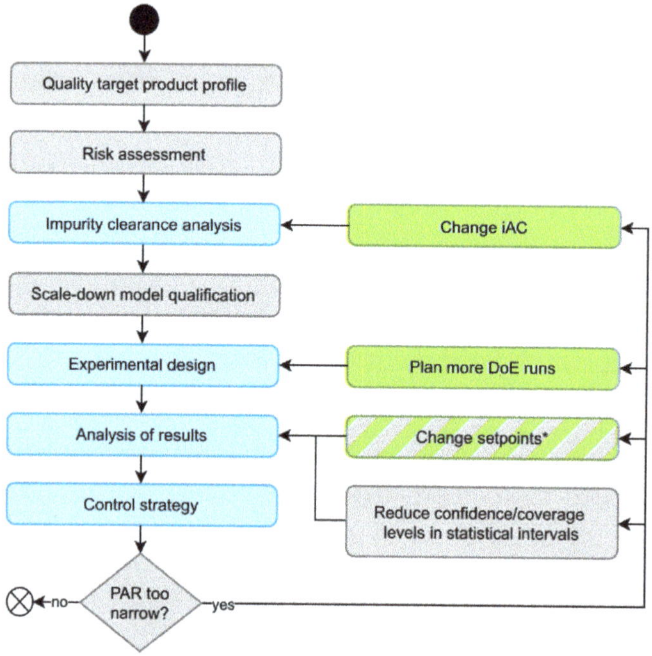

Figure 10. The left column represents common steps involved in process characterization, while the right column contains possible courses of action when no appropriate control strategy can be found. Green boxes are actions recommended by hDoE, and steps affected by hDoe are colored blue. * Changing PP setpoints could also be a viable, recommended action by future iterations of hDoE (see Section 4.2).

5. Conclusions

In this article, we introduced hDoE, an iterative tool for process development and characterization that facilitates a more effective way of gaining process understanding related to final product quality. This is essential, as it reduces experimental effort and time to market. To demonstrate the benefits of this approach quantitatively, we presented results from simulation studies where we chose the OOS rate as a measure of process understanding, which should be close to the OOS rate of the true (usually unknown) process. The benefit of this measure is that it includes both the mean and variability of the final product quality distribution as well as practically relevant limits (drug specifications). Other measures, such as the Kullback–Leibler divergence [30], also could have been used. However, we believe the OOS rate represents a more practically relevant measure and might be more tangible for process experts. We have demonstrated that hDoE leads to better overall process understanding with more than a 50% reduction in the number of experiments performed for simple scenarios. The reduction of experimental costs can even be increased for specific cases. hDoE starts out with an initial, minimal set of D-optimal runs on which the first set of unit operation regression models is fitted. Used as a recommender tool, either DoE or spiking runs are added in an iterative fashion guided by improvements in the predicted OOS rate. As the process is biased toward effects already found, we are using a recommender scheme akin to the one used in the Metropolis–Hastings algorithm to promote the detection of unknown effects and to improve parameter estimates. Of course, the overall quality of effect estimates is influenced by the number of runs available to the algorithm. However, our simulation studies show that a compromise between effects detected and runs invested can be found using a relatively low number of hDoE runs.

hDoE leverages the link between UOs as modeled by the IPM and thereby improves OOS rates by strategically recommending spiking studies at specific process steps. It provides valuable information to biopharmaceutical manufacturers about which type of experiment to perform next, and in which UO, and can decrease the time and money invested in experimental design. Potentially increased parameter uncertainty due to a lower number of samples is accounted for in the IPM's OOS prediction using conservative sampling and estimation methods such as tolerance intervals. We believe that hDoE is a viable option for experimental design that yields robust estimates of process properties while providing better control of the resources invested. This will lead to a substantial reduction of development costs and time to market, ultimately leading to more affordable drugs.

Author Contributions: Conceptualization, T.Z. and B.P.; methodology, T.Z. and T.O.; software, T.O. and B.P.; validation, B.P. and T.O.; formal analysis, T.O.; investigation, T.O.; data curation, T.O.; writing—original draft preparation, T.O. and T.Z.; writing—review and editing, T.O., T.Z., B.P. and C.H.; visualization, T.O.; supervision, C.H. and T.Z.; project administration, T.Z. All authors have read and agreed to the published version of the manuscript.

Funding: This work was supported by the Austrian Research Promotion Agency (FFG) (grant number: 844608) and within the framework of the Competence Center CHASE GmbH, funded by the Austrian Research Promotion Agency (grant number 868615) as part of the COMET program (Competence Centers for Excellent Technologies) by BMVIT, BMDW, and the Federal Provinces of Upper Austria and Vienna. Open Access Funding by TU Wien.

Institutional Review Board Statement: Not applicable.

Informed Consent Statement: Not applicable.

Data Availability Statement: All data generated or analyzed during this study are included in this published article.

Acknowledgments: This work was conducted within the COMET Centre CHASE, funded within the COMET—Competence Centers for Excellent Technologies program by the BMK, the BMDW and the Federal Provinces of Upper Austria and Vienna. The COMET program is managed by the Austrian Research Promotion Agency (FFG). The authors acknowledge TU Wien Bibliothek for financial support through its Open Access Funding Program.

Conflicts of Interest: The authors declare no conflict of interest.

References

1. ICH. *ICH Guideline Q8 (R2) on Pharmaceutical Development*; EMA: London, UK, 2017.
2. Burdick, R.; LeBlond, D.; Pfahler, L.; Quiroz, J.; Sidor, L.; Vukovinsky, K.; Zhang, L. *Statistical Applications for Chemistry, Manufacturing and Controls (CMC) in the Pharmaceutical Industry*; Springer: Cham, Switzerland, 2017.
3. FDA. *Process Validation: General Principles and Practices*; US FDA: Rockville, MD, USA, 2011.
4. Montgomery, D.C.; Peck, E.A.; Vining, G.G. *Introduction to Linear Regression Analysis*; John Wiley & Sons: Hoboken, NJ, USA, 2021.
5. Montgomery, D.C. *Design and Analysis of Experiments*; John Wiley & Sons: Hoboken, NJ, USA, 2017.
6. EMA. *Questions and Answers: Improving the Understanding of NORs, PARs, DSp and Normal Variability of Process Parameters*; EMA: London, UK, 2017.
7. Howe, W. Two-sided tolerance limits for normal populations—Some improvements. *J. Am. Stat. Assoc.* **1969**, *64*, 610–620.
8. Krishnamoorthy, K.; Mathew, T. *Statistical Tolerance Regions: Theory, Applications, and Computation*; John Wiley & Sons: Hoboken, NJ, USA, 2009.

9. Wallis, W.A. Tolerance intervals for linear regression. In Proceedings of the Second Berkeley Symposium on Mathematical Statistics and Probability, Berkeley, CA, USA, January 1951.
10. Francq, B.G.; Lin, D.; Hoyer, W. Confidence, prediction, and tolerance in linear mixed models. *Stat. Med.* **2019**, *38*, 5603–5622. [CrossRef]
11. Marschall, L.; Taylor, C.; Zahel, T.; Kunzelmann, M.; Wiedenmann, A.; Presser, B.; Studts, J.; Herwig, C. Specification-driven acceptance criteria for validation of biopharmaceutical processes. *Front. Bioeng. Biotechnol.* **2022**, *10*, 1010583. [CrossRef]
12. Seely, R.; Munyakazi, L.; Haury, J. Statistical tools for setting in-process acceptance criteria. *Dev. Biol.* **2003**, *113*, 17–26.
13. Wang, X.; Germansderfer, A.; Harms, J.; Rathore, A. Using statistical analysis for setting process validation acceptance criteria for biotech products. *Biotechnol. Prog.* **2007**, *23*, 55–60. [CrossRef] [PubMed]
14. Taylor, C.; Pretzner, B.; Zahel, T.; Herwig, C. Architectural & Technological Improvements to Integrated Bioprocess Models towards Real-Time Applications. *MDPI Bioeng.* **2022**, *9*, 534.
15. Darling, A. Considerations in performing virus spiking experiments and process validation studies. *Dev. Biol. Stand.* **1993**, *81*, 221–229. [PubMed]
16. Shukla, A.; Jiang, C.; Ma, J.; Rubacha, M.; Flansburg, L.; Lee, S. Demonstration of robust host cell protein clearance in biopharmaceutical downstream processes. *Biotechnol. Prog.* **2008**, *24*, 615–622. [CrossRef]
17. Johnson, R.; Montgomery, D.; Jones, B. An Expository Paper on Optimal Design. *Qual. Eng.* **2011**, *23*, 287–301. [CrossRef]
18. de Aguiar, F.; Bourguignon, B.; Khots, M.; Massart, D.; Phan-Than-Luu, R. D-optimal designs. *Chemom. Intell. Lab. Syst.* **1995**, *30*, 199–210. [CrossRef]
19. Goos, P.; Jones, B.; Syafitri, U. I-optimal design of mixture experiments. *J. Am. Stat. Assoc.* **2016**, *111*, 899–911. [CrossRef]
20. Jones, B.; Allen-Moyer, K.; Goos, P. A-optimal versus D-optimal design of screening experiments. *J. Qual. Technol.* **2021**, *53*, 369–382. [CrossRef]
21. Velayudhan, A. Overview of integrated models for bioprocess engineering. *Curr. Opin. Chem. Eng.* **2014**, *6*, 83–89. [CrossRef]
22. Mooney, C. *Monte Carlo Simulation*; Sage: Thousand Oaks, CA, USA, 1997.
23. Hahn, G. The hazards of extrapolation in regression analysis. *J. Qual. Technol.* **1977**, *9*, 159–165. [CrossRef]
24. Hamada, C.; Hamada, M. All-subsets regression under effect heredity restrictions for experimental designs with complex aliasing. *Qual. Reliab. Eng. Int.* **2010**, *26*, 75–81. [CrossRef]
25. Desboulets, L.D.D. A review on variable selection in regression analysis. *Econometrics* **2018**, *6*, 45. [CrossRef]
26. Hastings, W.K. *Monte Carlo Sampling Methods Using Markov Chains and Their Applications*; Oxford University Press: Oxford, UK, 1970.
27. Olusegun, A.M.; Dikko, H.G.; Gulumbe, S.U. Identifying the limitation of stepwise selection for variable selection in regression analysis. *Am. J. Theor. Appl. Stat.* **2015**, *4*, 414–419. [CrossRef]
28. Committee for Medicinal Products for Human Use. *Process Validation for the Manufacture of Biotechnology-Derived Active Substances and Data to Be Provided in Regulatory Submissions*; EMA: London, UK, 2016.
29. SAS Institute Inc. *JMP®16 Profilers*; SAS Institute Inc.: Cary, NC, USA, 2020–2021.
30. Joyce, J. Kullback-Leibler Divergence. In *International Encyclopedia of Statistical Science*; Springer: Berlin/Heidelberg, Germany, 2011; pp. 720–722.

bioengineering

MDPI

Article

Peristaltic Phenomenon in an Asymmetric Channel Subject to Inclined Magnetic Force and Porous Space

Muhammad Ijaz Khan, Maha M. A. Lashin, Nidhal Ben Khedher, Bilal Ahmed, Sami Ullah Khan and Mowffaq Oreijah et al.

1 Department of Mathematics and Statistics, Riphah International University, Islamabad 44000, Pakistan
2 College of Engineering, Princess Nourah Bint Abdulrahman University, Riyadh 11564, Saudi Arabia
3 Department of Mechanical Engineering, College of Engineering, University of Ha'il, Ha'il 81451, Saudi Arabia
4 Laboratory of Thermal and Energy Systems Studies, National School of Engineering of Monastir, University of Monastir, Monastir 5000, Tunisia
5 Department of Mathematics and Statistics, The University of Lahore, Sargodha Campusi, Punjab 40100, Pakistan
6 Department of Mathematics, COMSATS University Islamabad, Sahiwal 57000, Pakistan
7 Mechanical Engineering Department, College of Engineering and Islamic Architecture, Umm Al-Qura University, Makkah 21955, Saudi Arabia
8 Research Unity: Materials, Energy and Renewable Energies, Faculty of Science of Gafsa, University of Gafsa, Gafsa 2100, Tunisia
9 Faculty of Engineering and Technology, Future University in Egypt, New Cairo 11835, Egypt
10 Department of Mechanical Engineering, College of Engineering in Wadi Alddawasir, Prince Sattam Bin Abdulaziz University, Al-Kharj 16278, Saudi Arabia
11 Production Engineering and Mechanical Design Department, Faculty of Engineering, Mansoura University, Mansoura 35516, Egypt
* Correspondence: mikhan@math.qau.edu.pk

Citation: Muhammad Ijaz Khan, Maha M. A. Lashin, Nidhal Ben Khedher, Bilal Ahmed, Sami Ullah Khan and Mowffaq Oreijah et al. Peristaltic Phenomenon in an Asymmetric Channel Subject to Inclined Magnetic Force and Porous Space. *Bioengineering* **2022**, *9*, 588. https://doi.org/10.3390/bioengineering9100588

Academic Editors: Liang Luo and Giovanni Vozzi

Received: 30 August 2022
Accepted: 19 October 2022
Published: 20 October 2022

Abstract: This research is engaged to explore biological peristaltic transport under the action of an externally applied magnetic field passing through an asymmetric channel which is saturated with porous media. The set of governing partial differential equations for the present peristaltic flow are solved in the absence of a low Reynolds number and long wavelength assumptions. The governing equations are to be solved completely, so that inertial effects can be studied. The numerical simulations and results are obtained by the help of a finite element method based on quadratic six-noded triangular elements equipped with a Galerkin residual procedure. The inertial effects and effects of other pertinent parameters are discussed by plotting graphs based on a finite element (FEM) solution. Trapped bolus is discussed using the graphs of streamlines. The obtained results are also compared with the results given in the literature which are highly convergent. It is concluded that velocity and the number of boluses is enhanced by an increase in Hartmann number and porosity parameter K Increasing inertial forces increase the velocity of flow but increasing values of the porosity parameter lead to a decrease in the pressure gradient. The study elaborates that magnetic field and porosity are useful tools to control the velocity, pressure, and boluses in the peristaltic flow pattern.

Keywords: peristaltic flow; non-zero Reynolds number; asymmetric channel; finite element method; porous media

1. Introduction

The phenomenon in which fluid moves along the axis of a channel due to the contraction and relaxation of the walls is called peristalsis. Many scientists and engineers have been working on peristaltic flow due its interesting applications. Which involves discussion of Newtonian and non-Newtonian flows. The peristaltic flow of non-Newtonian fluids is currently an emerging field. Many applications of these phenomena have been seen in the engineering and bio-medical fields. Peristaltic flow also involved in human and animal organs, such as the flow of urine from the kidneys, blood flow to and from the heart, and the movement of chyme (partially digested food). Devices used in the printing industry, peristaltic pumps used in the chemical industry for manufacturing strong acids, and bio-medical devices work under the principle of peristalsis. Peristalsis is an emerging field which has attracted researchers to work on it. The first research took place in 1964, when Latham [1] worked on a peristaltic pump and demonstrated the peristaltic flow in the channel. He investigated peristalsis in a fixed frame of reference. After that, Shapiro et al. [2] studied the peristalsis phenomenon in a channel which was two-dimensional by geometry, but they used the assumptions of low lubrication theory which hides many inertial effects which should be discussed. They also presented theoretical results for both planes. Fung and Yih [3] investigated peristalsis in an infinite channel. In their investigation it was seen that peristaltic pumping produces reverse flow when its value is increased to a critical value, and they concluded that velocity profile is dependent on pressure gradient. They also observed that convergence of results decreases when the value of Reynolds numbers is too large. In an investigation by Hanin [4], large Reynolds numbers and a small wavelength were used. Zein and Ostrach [5] analyzed peristaltic phenomena for two-dimensional geometry with the limitations of long wavelength and low frequency. They obtained the solution in closed form and compared it with the result of Fung and Yih [3]. Peristaltic flow in tubes was also discussed by Barton and Raynor [6], but long-wave and short-wave disturbances were only discussed for Stokes flow. Chow [7] investigated peristaltic flow by employing long wavelength approximations. Peristaltic flow was also discussed in detail by Brown and Hung [8], who solved the governing equation with the help of a finite difference technique. Jaffrin [9] studied peristalsis in a two-dimensional tube by using a perturbation technique with the employment of non-linear effects. The detailed work on peristaltic phenomena in the channel by using the finite difference method was elaborated by Takabatake and Ayukawa [10]. They considered the complete set of equations that governs the problem. They examined the model with moderate values of Reynolds numbers and made a comparison with the results of perturbation analysis. An investigation on the full form of partial differential equations for peristaltic transport in the tube was also carried out by Takabatake et al. [11]. The trapping of bolus was also taken into account in their approach for studying peristaltic maxing. It is observed that pumping efficiency is greater in tubes in comparison with flat two-dimensional channels. Takabatake et al. [12] analyzed and studied the work which was undertaken earlier by Takabatake et al. [11] by using the finite element method (FEM), and compared the results with those obtained in the work of Dennis and Chang [13]. Takabatake [14] investigated pressure rise per wavelength by engaging the finite element method (FEM), and a comparison of the results with results built on perturbation analysis was also given in their study. Bhatti et al. [15] elaborated the peristaltic activity for Prandtl nanofluid transport with endoscopic annular geometry. Zeeshan et al. [16] focused on aspects of Jeffrey fluid flow within peristalsis. The determination of Prandtl nanoparticles following the peristaltic pattern was noticed in the framework of Abbasi et al. [17]. Riaz et al. [18] identified the thermal role of nano-sized particles with the slip flow problem due to peristaltic motion. Le et al. [19] pointed out the inertial consequences for the inclined channel flow due to peristaltic transport.

During the last few decades researchers and scientists have worked on peristaltic flow through porous media due to its wide range of application in various fields. Movement of fluid in plants and in the human body are examples of flow through porous media. Mining of rudimentary oil from the earth also involves a porous medium. What we mean by porous medium is something that includes minor holes over its whole surface which allow fluid to flow through it. Examples of porous media are sand, limestone, or a loaf of bakery bread. In living things such as humans, the gallbladder, kidney, and lungs are porous media. Hayat et al. [19] worked on peristaltic transport of Maxwell fluid which conducts electrically through a channel soaked with porous media. They found the solutions analytically and discussed the situations graphically. It was observed that mean velocity decreases with the increment of the Hall parameter. Srinivas and Kothandapani [20] found the analytical solution for peristaltic transport through a porous medium with the application of lubrication theory, which makes the governing equations very simple to discuss. It is seen that with an increase in the Schmidt number, the concentration decreases. Tripathi and Bég [21] investigated oscillating peristaltic flow of generalized Maxwell fluid through a porous medium by employing lubrication theory. They used the homotopy perturbation method to obtain the approximate analytical and numerical solutions. It was detected that the volumetric flow rate decreased when the relaxation time increased. Ellahi et al. [22] worked on the theoretical study of MHD effects on peristaltic flow in a rectangular duct through a porous medium employing long wavelength and low Reynolds number assumptions, but the solutions were obtained analytically. Trapping of bolus and the effects of MHD were also discussed. Sankad and Nagathan [23] studied MHD effects on peristaltic flow through porous media; the slip condition's effects were also taken into account and solutions were obtained analytically. The proposed results were discussed with the help of graphs. They modeled the governing equation by using low Reynolds number and long wavelength assumptions. It was observed that mean velocity rises with the magnetic field parameter in the presence of viscous damping force, but opposite behavior was seen when this force was removed.

In the present era, scientists are taking much interest in the discussion of fluid flowing through a channel or wall under the presence of an external magnetic force due to the number of applications in industry and astrophysics, etc., and its thought-provoking features. Magnetohydrodynamics (MHD) is very powerful tool in the medical field which is used in the diagnosis of hypothermia, tumors and MRI. The investigation of MHD peristaltic phenomena of a Newtonian fluid in a channel was presented by Yildirim and Sezer [24]. They obtained the solutions analytically and numerically by applying the homotopy perturbation method with low Reynolds number and long wavelength assumptions and made a comparison with the results of Nadeem and Akram [25]. It was seen that with an increase in L and M the temperature decreases. Nadeem and Akbar [26] studied MHD peristaltic flow of Newtonian fluid and discussed the effect of radially changing MHD. In the analysis, assumptions of lubrication theory were applied to obtained simplified governing equations, to facilitate ease of discussion. Manzoor el al. [27] discussed low-pressure plasma on degradation of multidrug resistant V. Cholera with antibacterial utilizations. Chu et al. [28] examined four different types of nanomaterials with thermal radiation and non-linear heat source and sink. Ramesh and Devaker [29] studied the peristaltic flow in porous asymmetric channel with the effects of external magnetic field and heat transfer. It is seen in the analysis that the pumping is increased at higher Hartmann numbers.

By the use of assumptions of low Reynolds numbers and long wavelength, the partial differential equations become very simple to discuss and hide many inertial effects which must be the part of discussion. To overcome this problem, Javed et al. [30] presented work on peristaltic flow in a two-dimensional inclined channel, employing an external magnetic field and without the use of lubrication theory. Many important inertial effects are discussed in their analysis; for example, they concluded that bolus size increased with increasing Reynolds numbers. The effects of Reynolds numbers on vorticity were also presented in their analysis. Ahmed at el. [31] also investigated peristaltic phenomena without employing low Reynolds number and long wavelength assumptions in a vertical channel. During analysis they concluded that heat generation controls the velocity, i.e., a rise in heat generation diminishes the velocity near the central region of the channel, and opposite behavior was observed near the walls. The peristaltic phenomena without the use of lubrication theory were also presented by Javed et al. [32]. They used the finite element method to obtain the numerical results. It was seen that the presented effects of Reynolds number are true for arbitrary values. Recently, some dynamic researchers investigated peristaltic flow [30–33] without engaging lubrication theory's assumptions about the channel, which is symmetric around its axis. In the literature, it is seen that many researchers have used lubrication theory, which leads the sets of partial differential equations that govern the peristaltic flow to be very simple to solve; however, those results leads to the exclusion of inertial terms in the discussion. As a result of this, the inertial effects and other significant aspects have not been studied. Some studies [34–38] interpret some basic concepts on fluid flow and nanomaterials.

The aim of the current investigation is to observe the peristaltic phenomenon in an asymmetric channel saturated with porous media, along with the applications of external orthogonal magnetic field effects without engaging the norms of low Reynolds numbers and long wavelengths, which causes the system to be non-linear. The governing equations will be solved in full form using the finite element method equipped with Galerkin's residual approach by using triangular elements in meshing with six nodes for numerical solution, which is highly convergent in comparison with linear triangular elements. Moreover, solving the system of equations in full form enables us to study the inertial effects and many other parameters on peristaltic flow. Besides this velocity distribution, pressure evaluation and trapping activity will be discussed with the help of graphs plotted based on numerical results.

2. Problem Formulation

In our present work we are considering an unsteady incompressible flow through a channel soaked with porous media. The considered flow is moving through a two-dimensional asymmetric channel of width $2a$ lying horizontally. Further, we have applied an external magnetic field of strength B_0 orthogonal to the direction of the flow. The fluid is a conductor electrically; due to this, an induced magnetic field arises, but it is insignificant in comparison with our applied B_0. Thus, we consider only the effect of the external magnetic field. The fluid flows due to the wave generated on the walls of the channel. The geometry of our considered problem is given in Figure 1. The upper moving wall of the infinite channel follows the relation given by [9,10]:

$$H_1(X,t) = a_1 + b_1\cos\left(\frac{2\pi(X-ct)}{\lambda}\right), \tag{1}$$

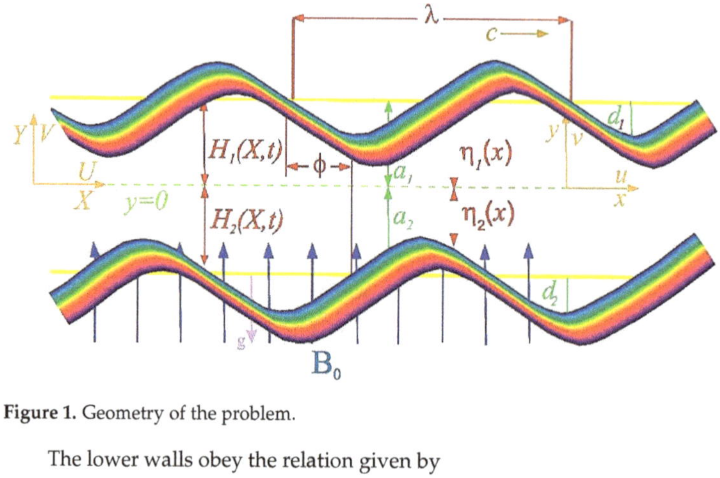

Figure 1. Geometry of the problem.

The lower walls obey the relation given by

$$H_2(X,t) = -a_2 - b_2\cos\left(\frac{2\pi(X-ct)}{\lambda} + \phi\right). \tag{2}$$

In the above relations, b_1 and b_2 represents the amplitudes, λ represents the wavelength of the wave, c is the speed of the wave, $2a$ is the width of the channel ($a_1 = a_2 = a$), and ϕ is the phase difference with ($0 < \phi \leq \pi$). Note that when $\phi = 0$ the corresponding channel is symmetric with the waves out of phase and when $\phi = \pi$ the corresponding channel is symmetric and the waves are in phase. Furthermore, a_1, a_2, b_1, b_2 and ϕ fulfil the requirements of the inlet of the divergent channel, i.e.,

$$a_1^2 + a_2^2 + 2b_1 b_2 \cos\phi \leq (a_1 + a_2)^2. \tag{3}$$

The governing equations for our present problem, i.e., the principles of conservation of momentum and conservation of mass and the law of for conducting fluid through a channel which is saturated with porous media under the magnetic field in fixed frame are expressed as [9,10]:

$$\nabla \times V^* = 0, \tag{4}$$

$$\rho(V^* \times \nabla V^*) = -\nabla P^* + \mu\left(\nabla^2 V^*\right) + (J \times B) - \frac{\mu}{k}V^*. \tag{5}$$

In the above equations, $V^* = (U, V, 0)$ is the velocity of fluid in two dimensions, and J and B are the current density and magnetic field, respectively. As the magnetic field is applied orthogonally, $B = (0,\ B_0,\ 0)$, ρ is the density, μ is the coefficient of dynamic viscosity, and k is the permeability of the porous media. The Lorentz force after neglecting the induced magnetic field (very small as compared with the external magnetic field) is given by

$$J \times B = \left[-\sigma B_0^2 U,\ 0,\ 0\right]. \tag{6}$$

The Equations (4) and (5) in a fixed frame take the form [9,10]:

$$\frac{\partial U}{\partial X} + \frac{\partial V}{\partial Y} = 0, \tag{7}$$

$$\rho\left\{\frac{\partial U}{\partial t} + U\frac{\partial U}{\partial X} + V\frac{\partial U}{\partial Y}\right\} = -\frac{\partial P}{\partial X} + \mu\left(\frac{\partial^2 U}{\partial X^2} + \frac{\partial^2 U}{\partial Y^2}\right) - \sigma B_0^2 U - \frac{\mu}{k}U, \tag{8}$$

$$\rho\left\{\frac{\partial V}{\partial t} + U\frac{\partial V}{\partial X} + V\frac{\partial V}{\partial Y}\right\} = -\frac{\partial P}{\partial Y} + \mu\left(\frac{\partial^2 V}{\partial X^2} + \frac{\partial^2 V}{\partial Y^2}\right) - \frac{\mu}{k}V, \tag{9}$$

with suitable boundary conditions given as

$$U = 0, \qquad V = \frac{\partial H_1}{\partial t}, \qquad at \qquad Y = H_1(X, t), \tag{10}$$

$$U = 0, \qquad V = \frac{\partial H_2}{\partial t}, \qquad at \qquad Y = H_2(X, t). \tag{11}$$

The conditions on U are due to no-slip conditions on the walls and the boundary conditions on V mean the wall velocity is same as the normal velocity of the walls. Now we transform our governing equations in a moving frame of reference $(x^*,\ y^*)$; for this, both frames of references are related, as

$$x^* = X - ct, \qquad y^* = Y, \qquad u^* = U - c, \qquad v^* = V, \qquad p^* = P, \tag{12}$$

where $(U,\ V)$ and (u^*, v^*) are velocity components in the laboratory frame and wave frame, respectively. p^* and P are pressures in the laboratory frame and wave frame. The governing equations after transformation in the moving frame using Equation (12) take the form:

$$\frac{\partial v^*}{\partial y^*} + \frac{\partial u^*}{\partial x^*} = 0, \tag{13}$$

$$\begin{aligned}\rho&\left\{v^* \tfrac{\partial u^*}{\partial y^*} + u^* \tfrac{\partial u^*}{\partial x^*}\right\} \\ &= -\tfrac{\partial p^*}{\partial x^*} + \mu\left(\tfrac{\partial^2 u^*}{\partial x^{*2}} + \tfrac{\partial^2 u^*}{\partial y^{*2}}\right) - \left(\sigma B_0^2 + \tfrac{\mu}{k}\right)(u^* + c),\end{aligned} \tag{14}$$

$$\rho\left\{\frac{\partial v^*}{\partial x^*}u^* + \frac{\partial v^*}{\partial y^*}v^*\right\} = -\frac{\partial p^*}{\partial y^*} + \mu\left(\frac{\partial^2 v^*}{\partial x^{*2}} + \frac{\partial^2 v^*}{\partial y^{*2}}\right) - \frac{\mu}{k}v^*, \tag{15}$$

with appropriate boundary conditions,

$$u^* = -c, v^* = \frac{2\pi c b_1}{\lambda}\sin\left(\frac{2\pi x^*}{\lambda}\right), \qquad at \quad y^* = \eta_1(x^*), \tag{16}$$

$$u^* = -c, v^* = -\frac{2\pi c b_2}{\lambda}\sin\left(\frac{2\pi x^*}{\lambda} + \phi\right), \qquad at \quad y^* = \eta_2(x^*), \tag{17}$$

where $\eta_1(x^*)$ represents the constraint on the upper peristaltic wall and $\eta_2(x^*)$ on the lower walls in the moving frame. The planes $y^* = \eta_1(x^*)$ and $y^* = \eta_2(x^*)$ follow the streamlines' pattern and the current flow rate q^* is constant throughout the cross section of the peristaltic channel in the moving reference frame. The consistent boundary conditions relating stream function are expressed as

$$\psi^* = q^*, \qquad at \quad y^* = \eta_1(x^*) \quad and \quad y^* = \eta_2(x^*). \tag{18}$$

ψ^* is the illustration of stream function and $q^* = Q^* - ca$ narrates the flow rates in the fixed frame and in the moving frame of reference. Two proposed dimensionless variables are given, by which we can make our differential equations dimensionless,

$$\begin{array}{llllll} x = \frac{x^*}{\lambda}, & y = \frac{y^*}{a_1}, & u = \frac{u^*}{c}, & v = \frac{v^*}{c}, & a = \frac{a_2}{a_1}, & b = \frac{b_1}{a_1}, \\ \overline{b} = \frac{b_2}{a_1}, & \eta_1 = \frac{\eta_1^*}{a_1}, & \eta_2 = \frac{\eta_2^*}{a_1}, & \alpha = \frac{a_1}{\lambda}, \; q = \frac{q^*}{ca_1}, & Re = \frac{ca_1}{\nu}\alpha, & \\ & M = \sqrt{\frac{\sigma}{\mu}}B_0 a_1, & K = \frac{k}{a_1^2}, & p = \frac{a_1^2 p^*(x^*)}{c\,\lambda\mu}. & & \end{array} \tag{19}$$

After making the dimensionless form of equations that governs the peristaltic phenomena and corresponding boundary conditions, we have

$$\alpha\frac{\partial u}{\partial x} + \frac{\partial v}{\partial y} = 0, \tag{20}$$

$$Re\left\{\alpha u\frac{\partial u}{\partial x} + v\frac{\partial u}{\partial y}\right\} = -\alpha\frac{\partial p}{\partial x} + \alpha\left(\alpha^2\frac{\partial^2 u}{\partial x^2} + \frac{\partial^2 u}{\partial y^2}\right) - \left(\alpha M^2 + \frac{\alpha}{K}\right)(u+1) \tag{21}$$

$$Re\left\{\alpha u\frac{\partial v}{\partial x} + v\frac{\partial v}{\partial y}\right\} = -\frac{\partial p}{\partial y} + \alpha\left(\alpha^2\frac{\partial^2 v}{\partial x^2} + \frac{\partial^2 v}{\partial y^2}\right) - \frac{\alpha}{K}v, \tag{22}$$

$$u = -1, v = 2\pi b\alpha\sin(2\pi x), \qquad at \quad y = \eta_1(x) = 1 + b\cos(2\pi x), \tag{23}$$

$$u = -1, v = -2\pi\overline{b}\,\alpha\sin(2\pi x + \phi), \quad at \quad y = \eta_2(x) = -a - \overline{b}\cos(2\pi x + \phi), \tag{24}$$

where Re is the ratio of inertial and viscous forces known as the Reynolds number, Fr represents the Forhheimer number ($Fr = \frac{c^2}{ga}$), and K is the permeability of the porous medium. After elimination of pressure terms by taking derivative Equation (21) with respect to y and Equation (22) with respect to x and simplifying by inserting the stream function and vorticity given by

$$u = \frac{\partial\psi}{\partial y}, \qquad v = -\alpha\frac{\partial\psi}{\partial x}, \qquad \omega = \left(\alpha\frac{\partial v}{\partial x} - \frac{\partial u}{\partial y}\right), \tag{25}$$

the governing equations in $\psi - \omega$ formulation take the form

$$\left(\alpha^2\frac{\partial^2\psi}{\partial x^2} + \frac{\partial^2\psi}{\partial y^2}\right) = -\omega, \tag{26}$$

$$Re\left\{\frac{\partial\psi}{\partial y}\frac{\partial\omega}{\partial x} - \frac{\partial\psi}{\partial x}\frac{\partial\omega}{\partial y}\right\} = \nabla^2\omega - \frac{\omega}{K} + M^2\frac{\partial^2\psi}{\partial y^2}. \tag{27}$$

The proposed boundary conditions in Equations (18), (23), and (24) become

$$\psi = q, \qquad \frac{\partial \psi}{\partial y} = -1, \qquad \frac{\partial \psi}{\partial x} = -2\pi b sin(2\pi x), \qquad at \;\; y = \eta_1(x), \tag{28}$$

$$\psi = q, \qquad \frac{\partial \psi}{\partial y} = -1, \qquad \frac{\partial \psi}{\partial x} = -2\pi \bar{b} sin(2\pi x + \phi), \qquad at \;\; y = \eta_2(x), \tag{29}$$

where $\nabla^2 = \alpha^2 \frac{\partial^2}{\partial x^2} + \frac{\partial^2}{\partial y^2}$. It is necessary to note that by employing assumptions of lubrication theory, Equation (26) vanishes and Equation (27) becomes

$$\frac{\partial^4 \psi}{\partial y^4} + M^2 \frac{\partial^4 \psi}{\partial y^4} = 0. \tag{30}$$

3. Finite Element Analysis

In this section we are going to obtain the numerical solution to partial differential equations that govern the problem. For this purpose the partial differential Equations (26) and (27) along with the proposed boundary conditions given in Equations (28) and (29) are considered to be solved by applying finite element method to obtained numerical solution. Using FEM, we can easily handle the complex domains and they have rapid convergence. The discretization technique used in the finite element method makes it more effective to solve non-linear problems. In our present problem, our aim is to solve the given governing equations in a comprehensive form so that the effects of Reynolds numbers and wave numbers can be discussed. For this, we discretize the domain into a non-uniform mesh based on triangular elements (quadratic) founded on six nodes using pdetool in MATLAB. Firstly, we find the solution on every element of the domain, then assemble all elements into a large matrix known as the global matrix. The resultant equations are a non-linear system of algebraic equations. Then we will use the Newton Raphson method to obtain the solution to the equations. The finite element method will be implemented as follows:

The nodal variables are approximated as

$$\psi = \sum_{k=1}^{n} N_k \psi_k, \qquad \omega = \sum_{k=1}^{n} N_k \omega_k, \tag{31}$$

where ψ_k and ω_k denote the nodal approximation of ψ and ω, respectively. In the subsequent use of the finite element method, the weighted residuals of Equations (26) and (27) can be written as

$$R_i^1 = \int_\Omega w_1 \left(\alpha^2 \frac{\partial^2 \psi}{\partial x^2} + \frac{\partial^2 \psi}{\partial y^2} + \omega \right) d\Omega, \tag{32}$$

$$R_i^2 = \int_\Omega w_2 \left[Re \left(\frac{\partial \psi}{\partial y} \frac{\partial \omega}{\partial x} - \frac{\partial \psi}{\partial x} \frac{\partial \omega}{\partial y} \right) - \left(\alpha^2 \frac{\partial^2 \omega}{\partial x^2} + \frac{\partial^2 \omega}{\partial y^2} \right) + \frac{\omega}{K} - M^2 \frac{\partial^2 \psi}{\partial y^2} \right] d\Omega, \tag{33}$$

where w_1 and w_2 are weight functions and $d\Omega = d\psi d\omega$. Simplifying Equations (32) and (33), the weak formulation yields the following equations:

$$R_i^1 = \int_\Omega \left(\alpha^2 \frac{\partial \psi}{\partial x} \frac{\partial w_1}{\partial x} + \frac{\partial \psi}{\partial y} \frac{\partial w_1}{\partial y} \right) d\Omega - \int_\Omega w_1 \omega \, d\Omega - \int_\Omega w_1 \frac{\partial \psi}{\partial n} \, d\Gamma, \tag{34}$$

$$\begin{gathered} R_i^2 = Re \int_\Omega w_2 \left(\frac{\partial \psi}{\partial y} \frac{\partial \omega}{\partial x} - \frac{\partial \psi}{\partial x} \frac{\partial \omega}{\partial y} \right) d\Omega + \int_\Omega \left(\alpha^2 \frac{\partial \omega}{\partial x} \frac{\partial w_2}{\partial x} + \frac{\partial \omega}{\partial y} \frac{\partial w_2}{\partial y} \right) d\Omega + \int_\Omega w_2 \frac{\omega}{K} d\Omega \\ + M^2 \int_\Omega \frac{\partial \psi}{\partial y} \frac{\partial w_2}{\partial y} d + \int_\Gamma w_2 \frac{\partial \omega}{\partial n} \, d\Gamma - M^2 \int_\Gamma w_2 \frac{\partial \psi}{\partial n} \, d\Gamma, \end{gathered} \tag{35}$$

where Ω denotes the the area integral for every element and Γ is the boundary for every element. Now, Re = 5 using Equation (31) in Equations (34) and (35), we get

$$R_i^1 = \sum_{i=1}^{n} \int_\Omega \left(\alpha^2 \frac{\partial N_i}{\partial x}\frac{\partial N_K}{\partial x} + \frac{\partial N_i}{\partial y}\frac{\partial N_K}{\partial y}\right)\psi_i d\Omega - \sum_{i=1}^{n} \int_\Omega N_i N_k \omega_i d\Omega - \int_\Gamma N_k d\Gamma \tag{36}$$

$$\begin{aligned} R_i^2 = {} & Re \sum_{i=1}^{n} \int_\Omega N_k \left(\tfrac{\partial N_i}{\partial y}\tfrac{\partial N_i}{\partial x} + \tfrac{\partial N_i}{\partial x}\tfrac{\partial N_i}{\partial y}\right)\psi_i \omega_i d\Omega \\ & + \sum_{i=1}^{n} \int_\Omega \left(\alpha^2 \tfrac{\partial N_i}{\partial x}\tfrac{\partial N_K}{\partial x} + \tfrac{\partial N_i}{\partial y}\tfrac{\partial N_K}{\partial y}\right)\omega_i d\Omega + \tfrac{1}{K}\sum_{i=1}^{n} \int_\Omega N_k N_i \omega_i d\Omega \\ & + M^2 \sum_{i=1}^{n} \int_\Omega \tfrac{\partial N_i}{\partial y}\tfrac{\partial N_K}{\partial y} d\Omega + \int_\Gamma N_k d\Gamma - M^2 \int_\Gamma N_k d\Gamma \end{aligned} \tag{37}$$

The assembled global matrix takes the form

$$KA = F. \tag{38}$$

Equation (38) leads to a system of algebraic equations which occur non-linearly and can be solved by any iterative method. In the present investigation the well-known Newton Raphson method is used to solve the given system. The elaborated procedure is iterative to the desired convergence rate.

Peristaltic waves produced on the walls induce periodic flow. Due to this, we can evaluate pressure only on the dominant middle part of the channel, which is equal to one wavelength. To evaluate pressure, the expression in dimensionless form is given by

$$\frac{\partial p}{\partial x} = Re\left\{\frac{\partial^2 \psi}{\partial y^2}\frac{\partial \psi}{\partial x} - \frac{\partial^2 \psi}{\partial x \partial y}\frac{\partial \psi}{\partial y}\right\} - \frac{\partial \omega}{\partial y} - \left(M^2 + \frac{1}{K}\right)\left(\frac{\partial \psi}{\partial y} + 1\right), \tag{39}$$

$$\frac{\partial p}{\partial y} = \alpha^2 Re\left\{\frac{\partial^2 \psi}{\partial x^2}\frac{\partial \psi}{\partial y} - \frac{\partial^2 \psi}{\partial x \partial y}\frac{\partial \psi}{\partial x}\right\} + \alpha^2\frac{\partial \omega}{\partial x} + \frac{\alpha^2}{K}\frac{\partial \psi}{\partial x}. \tag{40}$$

The pressure rise in the moving frame of reference is well defined as

$$\Delta P_\lambda = \int_0^\lambda \frac{dp}{dx} dx. \tag{41}$$

4. Validation of Numerical Results

This section is to check the validity of our present investigation by making a comparison between our obtained numerical results and the result of Mishra and Rao [33] in the limiting case, i.e., using the assumptions of lubrication theory in our present investigation, that is, $Re = 0$, $\alpha = 0$, $M = 0$, $d_1 = 0$, $d_2 = 0$, and $\gamma = 0$. The obtained solutions in the investigation by Mishra and Rao [33] are analytic. The velocity distribution based on our present numerical solutions is compared with that of Mishra and Rao [33] by using plots. It can be clearly observed from Figure 2 that in the limit of $\gamma = 0, M = 0$ and $d_1 = 0$, $d_2 = 0$, our calculated results have a remarkably good convergence rate with the results of Mishra and Rao [33]. Thus, it is indicated that the current study will be very useful for further research.

Analysis of the numerical results is presented in this section. Plots of velocity, streamlines, and pressure gradient are obtained and elaborated. Detailed discussion of the trapping of bolus is also presented. The effect of other pertinent parameters such as Reynolds number Re, Hartman number M, wave number α, and phase difference ϕ is also described.

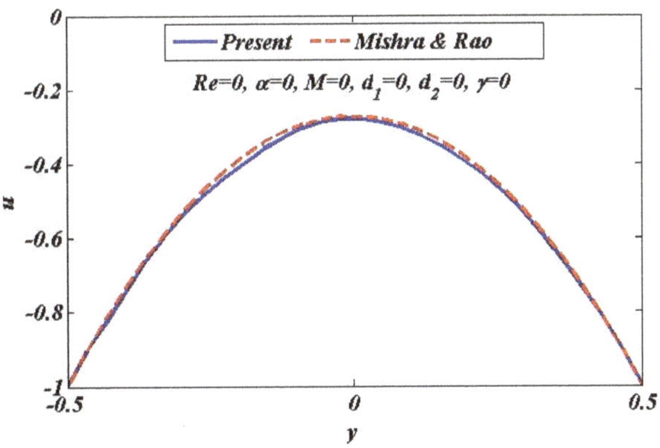

Figure 2. Comparative analysis for results with Mishra and Rao [33] for velocity distribution in limitation of $Re = 0$, $\alpha = 0$, $M = 0$, $d_1 = 0$, $d_2 = 0$, and $\gamma = 0$.

4.1. Velocity Profile

This subsection is presented here to discuss the behavior of velocity of fluid flowing through an asymmetric channel peristaltically. For this reason, a graph of longitudinal-velocity at cross section $x = 0$ is plotted for various changing values of mean flow rate Q, Reynolds number Re, Hartmann Number M, porosity of porous media K, and wave number α in Figures 3–7. From these plots it can be clearly concluded that the velocity of fluid on the walls of the channel vanishes due to no-slip conditions on the boundary. Figure 3 is designed to scrutinize the vibrational effects of mean flow rate Q for fixed values of other involved parameters. Time-mean flow rate measures the quantity of fluid passing through cross-sectional area of the channel. As we increase the value of time-mean flow rate Q, it means that more fluid is moving through the cross section, which means that the velocity is going to be increased. Figure 3 illustrates that increasing values of mean flow rate with rapid augmentation in the velocity profile are observed in the middle part and at the sides. The Reynolds number can be clearly defined as the ratio between inertial and viscous forces, so enhancing the value of Re leads to a decrease in viscous forces, which result in a decrease in opposition offered to the fluid. It can be seen from Figure 4 that by enhancing the value of the Reynolds number the longitudinal velocity is enhanced near the upper wall of the peristaltic channel, but reverse behavior is seen near the lower wall and mixed performance is seen in the middle part of the channel. Figure 5 is plotted to elaborate the effects of wave number for diverse values of α with static values of other parameters, i.e., $Re = 1,\ Q = 1.5,\ M = 2,\ d_1 = 0.3,\ d_2 = 0.3,\ K = 0.1$, and $\phi = \pi$. It shows that an increase in α results in a decline in the velocity near to the upper wall, and an increase in velocity of fluid is observed near the lower wall. Figures 6 and 7 elaborate the changing effects of the Hartmann number M and porosity parameter K by keeping the other parameters fixed. It is evident that with an enhancement in the porosity parameter and magnetic field, an increase in velocity is observed in the central part of the peristaltic channel but velocity decreases near the walls. It may be concluded from Figures 6 and 7 that the Hartmann number and porosity parameter are the best way to control the velocity of peristaltic flow.

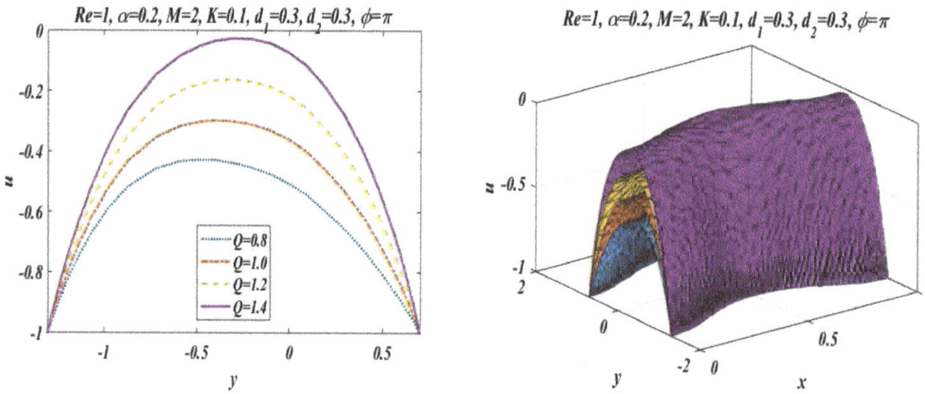

Figure 3. Velocity change for Q with $Re = 1$, $\alpha = 0.2$, $M = 2$, $K = 0.1$, $d_2 = 0.3$, $d_1 = 0.3$ and $\phi = \pi$.

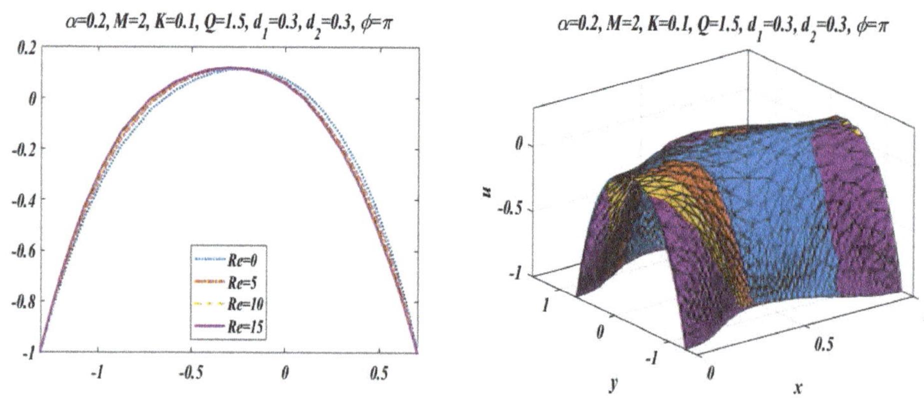

Figure 4. Variation in velocity for Re with $Q = 1.5$, $\alpha = 0.2$, $M = 2$, $K = 0.1$, $d_1 = 0.3$, $d_2 = 0.3$ and $\phi = \pi$.

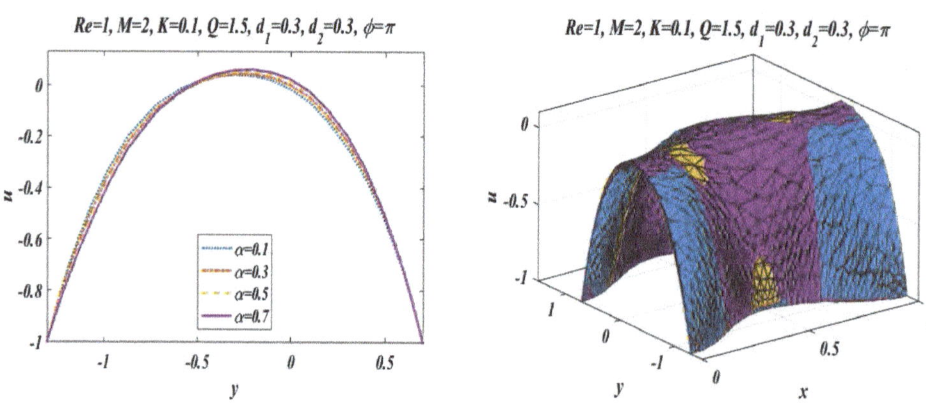

Figure 5. Velocity profile for α with $Re = 1$, $Q = 1.5$, $M = 2$, $K = 0.1$, $d_1 = 0.3$, $d_2 = 0.3$ and $\phi = \pi$.

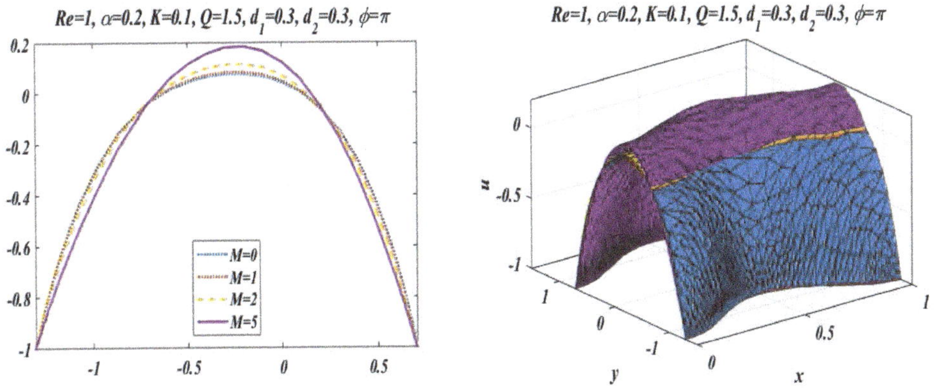

Figure 6. Velocity profile for M with $Re = 1$, $\alpha = 0.2$, $Q = 1.5$, $K = 0.1$, $d_1 = 0.3$, $d_2 = 0.3$ and $\phi = \pi$.

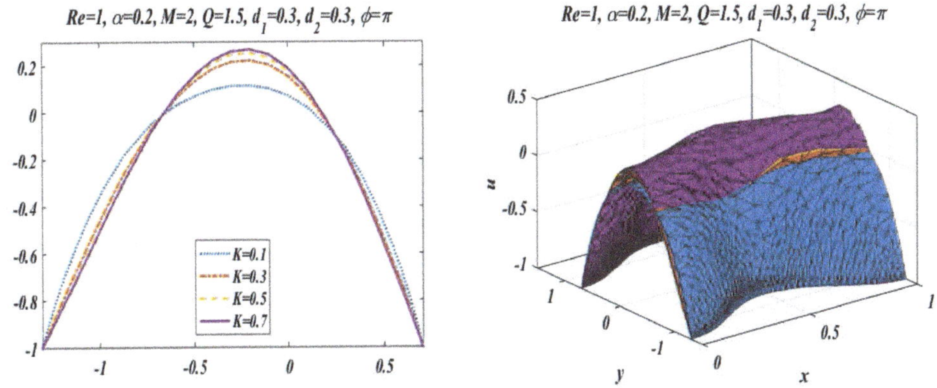

Figure 7. Velocity profile for K with $Re = 1$, $\alpha = 0.2$, $M = 2$, $Q = 1.5$, $d_1 = 0.3$, $d_2 = 0.3$ and $\phi = \pi$.

4.2. Pressure Distribution

In the peristaltic phenomena, the fluid flows due to the waves which are produced on the walls of the channel. These waves produce a pressure which causes fluid to flow forward. This subsection is devoted to discussing the effect of involved pertinent parameters on pressure-rise per wavelength by changing the value of one parameter while keeping the others fixed. The variation effects of the Hartmann number M, Reynolds number Re, wave number α, porosity parameter K, and phase difference ϕ on pressure gradient is particularized in Figures 8–12 by plotting the graph in the positive pumping region only. It is perceived that pressure increases with a rise in Hartmann number M, wave number, Reynolds number, and phase difference; additionally, a lessening in pressure is predicted in the augmented region. In contrast to the Hartmann and wave numbers, it is seen that pressure rise per wavelength declines with increase in the value of porosity parameter K in the peristaltic pumping region.

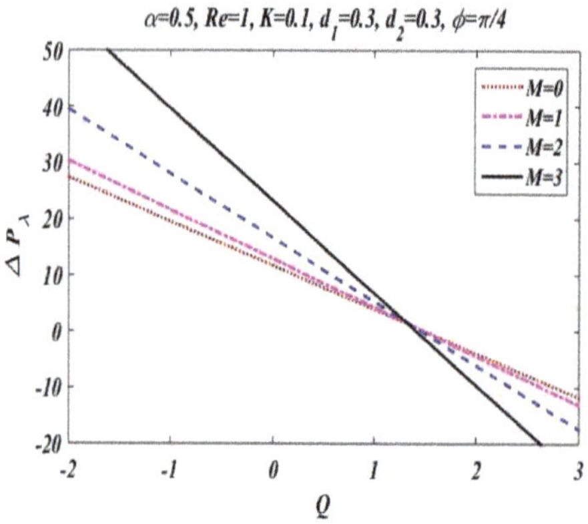

Figure 8. Pressure gradient for M.

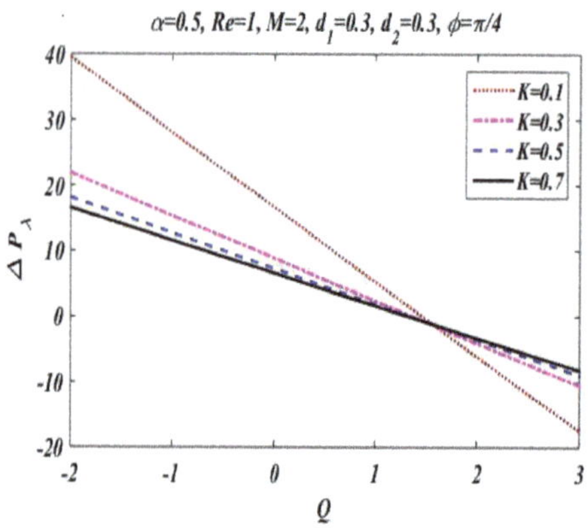

Figure 9. Pressure gradient for K.

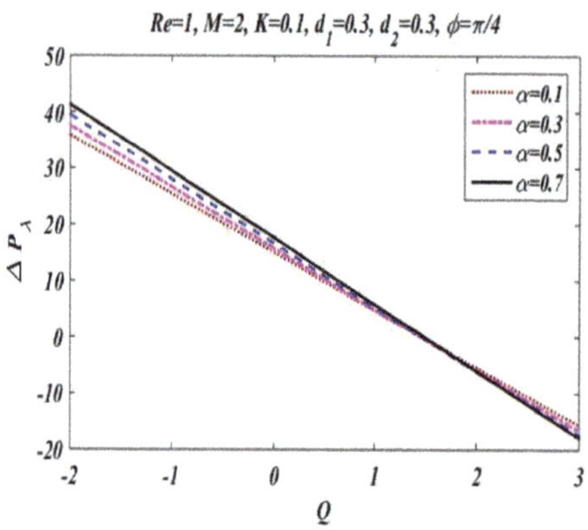

Figure 10. Pressure gradient for α.

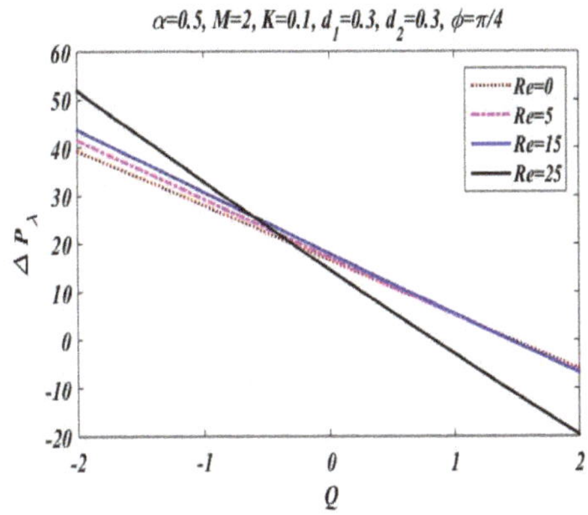

Figure 11. Pressure gradient for *Re*.

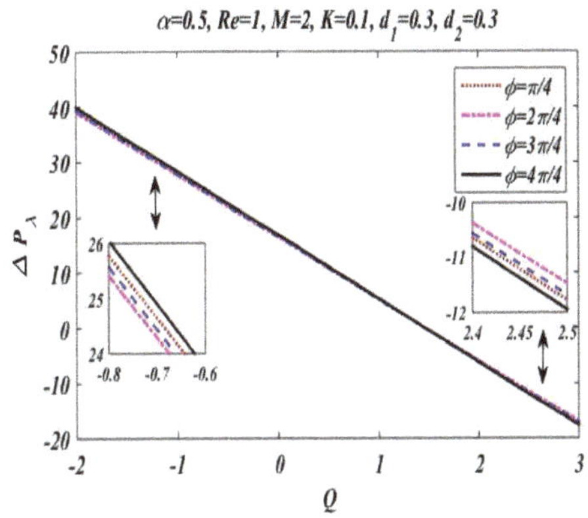

Figure 12. Pressure gradient for ϕ.

4.3. Trapping of Bolus

The graph of streamlines is plotted in this subsection to describe the useful trapping phenomenon for different of values of Reynolds number Re, Hartmann number M, and porosity of porous media K for fixed values of other parameters with changing values of phase difference. Streamlines describes how the fluid is moving physically. Smooth streamlines correspond to smooth flow. As waves are generated on the wavy walls, these streamlines become more curved and this leads to enclosing a bolus which moves with the flow. In Figure 13, streamlines are plotted with an unlike phase shift for $Re = 1$ and $Re = 5$ with static values of other related parameters, i.e., $\alpha = 0.2, K = 0.1,\ M = 1,\ d_1 = d_2 = 0.5$, and $\mathrm{Q} = 1.6$. It is noted that when $\phi = 0$, it leads to symmetric channel, and there is no effect on bolus size by enhancing the value of the Reynolds number. However, when $\phi = \pi/4, 2\pi/4, 3\pi/4$, this leads to an asymmetric channel. It should be noted that there is a slight change in bolus size, but as the asymmetry of the channel increases, it cause less retardation in the flow. Figures 14 and 15 show that increasing the values of the applied magnetic field and porosity parameter leads to an increase in the size and number of boluses for both symmetric and asymmetric channels, which causes more resistance and decreases the fluid flow. It is also observed that in case of enhancement of K, the number of boluses is more than in the case of an increase in the Hartmann number M.

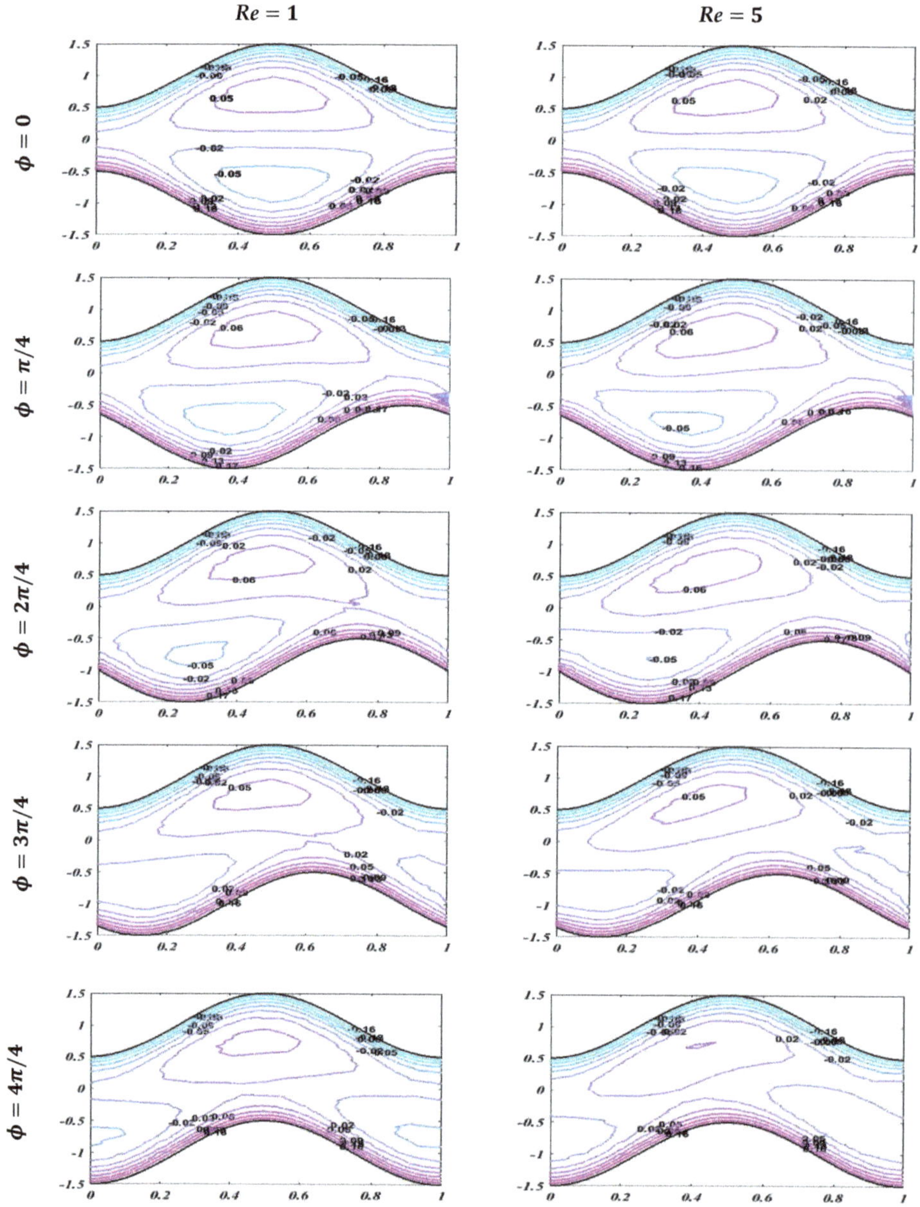

Figure 13. Streamlines with diverse phase shift for $Re = 1$ and $Re = 5$ with static values of other involved variables $\alpha = 0.2$, $K = 0.1$, $M = 1$, $d_1 = d_2 = 0.5$, and $Q = 1.6$.

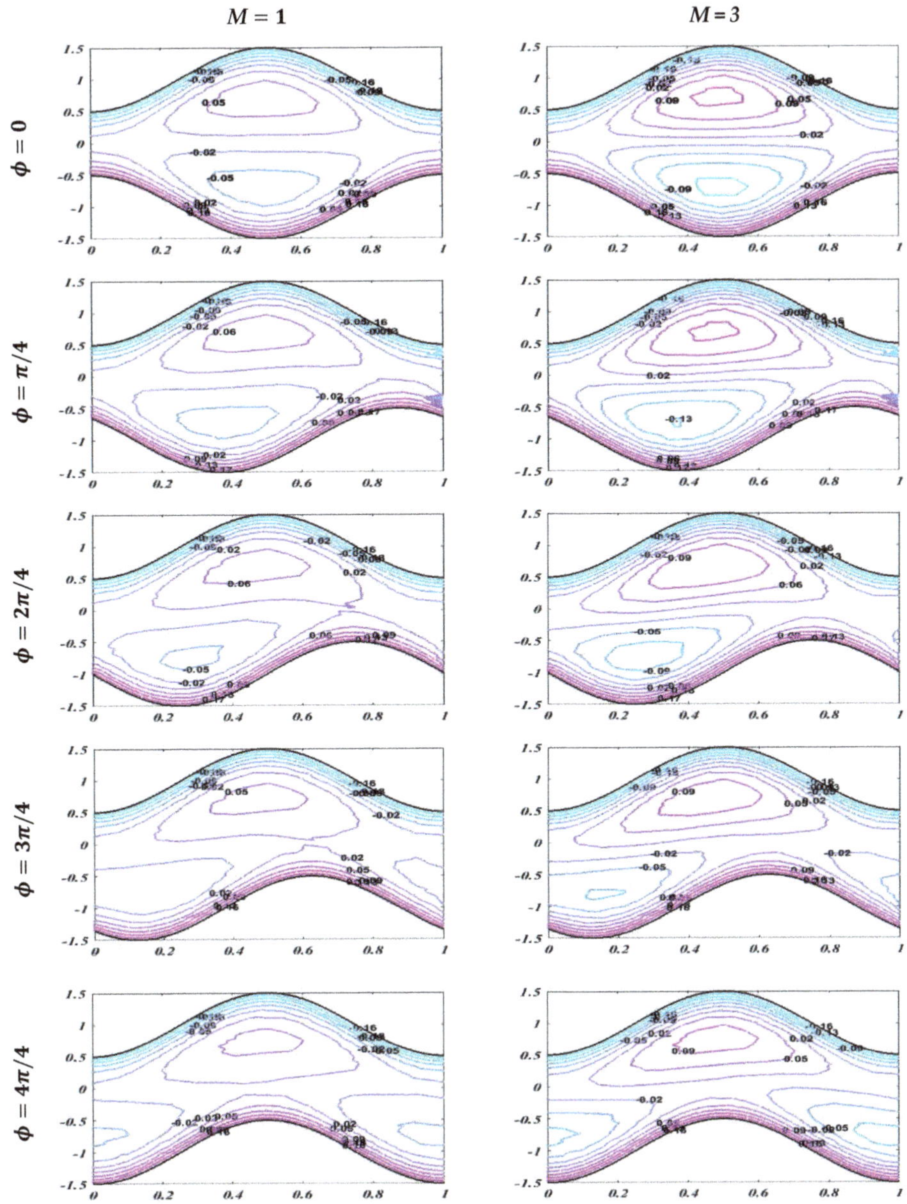

Figure 14. Streamlines with diverse Hartmann numbers for $M = 1$ and $M = 3$ with static values of other involved variables $\alpha = 0.2$, $Re = 1$, $K = 0.1$, $d_1 = d_2 = 0.5$, and $Q = 1.6$.

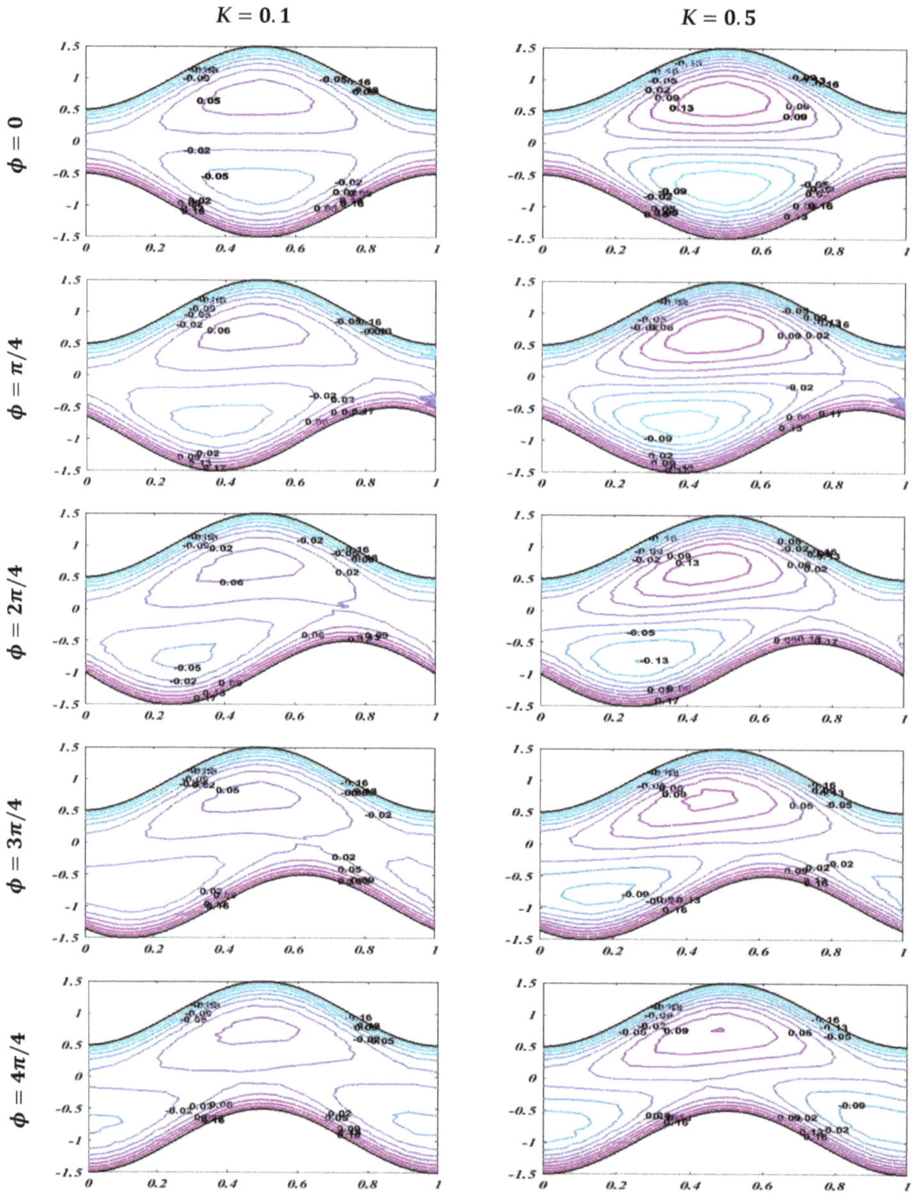

Figure 15. Streamlines with $K = 0.1$ and $K = 0.3$ when $\alpha = 0.2$, $Re = 1$, $M = 1$, $d_1 = d_2 = 0.5$, and $Q = 1.6$.

5. Conclusions

In the present work the investigation of peristaltic transport is carried out in an asymmetric channel under an applied magnetic field. The numerical simulations are carried out using the finite element method. Moreover, lubrication theory is not used in this study. The present work leads to the following conclusions:

The velocity is enhanced by increasing time-mean flow rate Q and Reynolds number Re but declines with an increase in the value of wave number α.

It is seen that by increasing the Hartmann number and porosity parameter K, the velocity of peristaltic flow is enhanced, but pressure distribution declines with increasing porosity parameter.

Increasing the value of the Reynolds number leads to enhancement in the velocity and pressure of the flow. However, there is a slight change in the size of the bolus.

The streamlines illustrate that the number of boluses is increased by enhancing the Hartmann number and porosity parameter.

It is concluded that magnetic field and porosity are the best factors with which to control the velocity, pressure, and boluses in the peristaltic flow.

Author Contributions: Conceptualization, B.A. and S.U.K.; methodology, B.A.; software, B.A.; validation, M.I.K., M.M.A.L. and A.M.G.; formal analysis, K.G.; investigation, M.O.; resources, E.S.M.T.-E.; data curation, S.U.K.; writing—original draft preparation, B.A. and M.I.K.; writing—review and editing, N.B.K., M.O. and K.G.; visualization, E.S.M.T.-E. and A.M.G.; supervision, S.U.K. and B.A.; project administration, E.S.M.T.-E. and M.I.K. All authors have read and agreed to the published version of the manuscript.

Funding: This research received no external funding.

Institutional Review Board Statement: Not applicable.

Informed Consent Statement: Not applicable.

Data Availability Statement: All the data are clearly mentioned in the manuscript.

Acknowledgments: The authors would like to thank the Deanship of Scientific Research at Umm Al-Qura University for supporting this work by Grant Code: (22UQU4331317DSR84); The authors express their grat-itude to Princess Nourah bint Abdulrahman University Researchers Supporting Project (Grant No. PNURSP2022R152), Princess Nourah bint Abdulrahman University, Riyadh, Saudi Arabia.

Conflicts of Interest: The authors declare no conflict of interest.

References

1. Latham, T.W. Fluid Motions in a Peristaltic Pump. Ph.D. Thesis, Massachusetts Institute of Technology, Cambridge, MA, USA, 1966.
2. Shapiro, A.H.; Jaffrin, M.Y.; Weinberg, S.L. Peristaltic pumping with long wavelengths at low Reynolds number. *J. Fluid Mech.* **1969**, *37*, 799–825. [CrossRef]
3. Fung, Y.C.; Yih, C.S. Peristaltic transport. *J. Appl. Mech.* **1968**, *35*, 669–675. [CrossRef]
4. Hanin, M. The flow through a channel due to transversally oscillating walls (Mean flow rate calculated for flow in two dimensional channel generated by transverse deflection oscillations along walls). *Isr. J. Technol.* **1968**, *6*, 67–71.
5. Zien, T.F.; Ostrach, S. A long wave approximation to peristaltic motion. *J. Biomech.* **1970**, *3*, 63–75. [CrossRef]
6. Barton, C.; Raynor, S. Peristaltic flow in tubes. *Bull. Math. Biophys.* **1968**, *30*, 663–680. [CrossRef] [PubMed]
7. Chow, T.S. Peristaltic transport in a circular cylindrical pipes. *J. Appl. Mech.* **1970**, *37*, 901–905. [CrossRef]
8. Brown, T.D.; Hung, T.K. Computational and experimental investigations of two-dimensional nonlinear peristaltic flows. *J. Fluid Mech.* **1977**, *83*, 249–272. [CrossRef]
9. Jaffrin, M.Y. Inertia and streamline curvature effects on peristaltic pumping. *Int. J. Eng. Sci.* **1973**, *11*, 681–699. [CrossRef]
10. Takabatake, S.; Ayukawa, K. Numerical study of two-dimensional peristaltic flows. *J. Fluid Mech.* **1982**, *122*, 439–465. [CrossRef]
11. Takabatake, S.; Ayukawa, K.; Mori, A. Peristaltic pumping in circular cylindrical tubes: A numerical study of fluid transport and its efficiency. *J. Fluid Mech.* **1988**, *193*, 267–283. [CrossRef]
12. Takabatake, S.; Ayukawa, K.; Sawa, M. Finite-element analysis of two-dimensional peristaltic flows: 1st report, finite-element solutions. *JSME Int. J. Bull. JSME* **1987**, *30*, 2048–2049. [CrossRef]
13. Dennis, S.C.R.; Chang, G.Z. Numerical Integration of the Navier-Stokes Equations for Steady Two-Dimensional Flow. *Phys. Fluids* **1969**, *12*, II-88. [CrossRef]
14. Takabatake, S.; Ayukawa, K.; Sawa, M. Finite element analysis of two-dimensional peristaltic flow (2nd report, pressure-flow characteristics). *Jpn. Soc. Mechanical. Eng.* **1990**, *56*, 3633–3637.
15. Bhatti, M.M.; Zeeshan, A.; Ellahi, R.; AnwarBég, O.; Kadir, A. Effects of coagulation on the two-phase peristaltic pumping of magnetized prandtl biofluid through an endoscopic annular geometry containing a porous medium. *Chin. J. Phys.* **2019**, *58*, 222–234. [CrossRef]
16. Zeeshan, A.; Ijaz, N.; Abbas, T.; Ellahi, R. The Sustainable Characteristic of Bio-Bi-Phase Flow of Peristaltic Transport of MHD Jeffrey Fluid in the Human Body. *Sustainability* **2018**, *10*, 2671. [CrossRef]

17. Abbasi, A.; Mabood, F.; Farooq, W.; Khan, S.U. Radiation and joule heating effects on electroosmosis-modulated peristaltic flow of Prandtl nanofluid via tapered channel. *Int. Commun. Heat Mass Transf.* **2021**, *123*, 105183. [CrossRef]
18. Riaz, A.; Khan, S.U.; Zeeshan, A.; Khan, S.U.; Hassan, M.; Muhammad, T. Thermal analysis of peristaltic flow of nanosized particles within a curved channel with second-order partial slip and porous medium. *J. Therm. Anal. Calorim.* **2021**, *143*, 1997–2009. [CrossRef]
19. Ge-JiLe, H.; Ahmed, B.; Al-Khaled, K.; Mehdi, M.T.; Khan, S.U.; Khan, M.I.; Chu, Y. Peristaltic activity in an asymmetric inclined channel with inertial forces under the inducement of magnetic field: Finite Element Method. *Alex. Eng. J.* **2021**, *60*, 4723–4734. [CrossRef]
20. Srinivas, S.; Kothandapani, M. The influence of heat and mass transfer on MHD peristaltic flow through a porous space with compliant walls. *Appl. Math. Comput.* **2009**, *213*, 197–208. [CrossRef]
21. Tripathi, D.; Bég, O.A. A numerical study of oscillating peristaltic flow of generalized Maxwell viscoelastic fluids through a porous medium. *Transp. Porous Media* **2012**, *95*, 337–348. [CrossRef]
22. Ellahi, R.; Bhatti, M.M.; Riaz, A.; Sheikholeslami, M. Effects of magnetohydrodynamics on peristaltic flow of Jeffrey fluid in a rectangular duct through a porous medium. *J. Porous Media* **2014**, *17*, 143–157. [CrossRef]
23. Sankad, G.C.; Nagathan, P.S. Unsteady MHD peristaltic flow of a couple stress fluid through porous medium with wall and slip effects. *Alex. Eng. J.* **2016**, *55*, 2099–2105. [CrossRef]
24. Yıldırım, A.; Sezer, S.A. Effects of partial slip on the peristaltic flow of a MHD Newtonian fluid in an asymmetric channel. *Math. Comput. Model.* **2010**, *52*, 618–625. [CrossRef]
25. Nadeem, S.; Akram, S. Heat transfer in a peristaltic flow of MHD fluid with partial slip. *Commun. Nonlinear Sci. Numer. Simul.* **2010**, *15*, 312–321. [CrossRef]
26. Nadeem, S.; Akbar, N.S. Influence of radially varying MHD on the peristaltic flow in an annulus with heat and mass transfer. *J. Taiwan Inst. Chem. Eng.* **2010**, *41*, 286–294. [CrossRef]
27. Manzoor, N.; Qasim, I.; Khan, M.I.; Ahmed, M.W.; Guedri, K.; Bafakeeh, O.T.; Eldin, E.S.M.T.; Galal, A.M. Antibacterial applications of low-pressure plasma on degradation of multidrug resistant. *V. Cholera. Appl. Sci.* **2022**, *12*, 9737. [CrossRef]
28. Chu, Y.M.; Khan, M.I.; Abbas, T.; Sidi, M.O.; Alharbi, K.A.M.; Alqsair, U.F.; Khan, S.U.; Khan, M.R.; Malik, M.Y. Radiative thermal analysis for four types of hybrid nanoparticles subject to non-uniform heat source: Keller Box numerical approach. *Case Stud. Therm. Eng.* **2022**, *40*, 102474. [CrossRef]
29. Ramesh, K.; Devakar, M. Magnetohydrodynamic peristaltic transport of couple stress fluid through porous medium in an inclined asymmetric channel with heat transfer. *J. Magn. Magn. Mater.* **2015**, *394*, 335–348. [CrossRef]
30. Javed, T.; Hamid, A.H.; Ahmed, B.; Ali, N. Effect of high Reynolds number on hydromagnetic peristaltic flow in an inclined channel using finite element method. *J. Korean Phys. Soc.* **2017**, *71*, 950–962. [CrossRef]
31. Ahmed, B.; Javed, T.; Hamid, A.H.; Sajid, M. Numerical analysis of mixed convective peristaltic flow in a vertical channel in presence of heat generation without using lubrication theory. *J. Appl. Fluid Mech.* **2017**, *10*, 1813–1827. [CrossRef]
32. Javed, T.; Ahmed, B.; Ali, N.; Hamid, A.H. Finite element analysis of the hydromagnetic peristaltic flow in a porous-saturated channel at moderate Reynolds numbers. *J. Porous Media* **2017**, *20*, 841–857. [CrossRef]
33. Mishra, M.; Rao, A.R. Peristaltic transport of a Newtonian fluid in an asymmetric channel. *Z. Und Phys. ZAMP* **2003**, *54*, 532–550. [CrossRef]
34. Abbasi, A.; Farooq, W.; Tag-ElDin, E.S.M.; Khan, S.U.; Khan, M.I.; Guedri, K.; Elattar, S.; Waqas, M.; Galal, A.M. Heat transport exploration for hybrid nanoparticle (Cu, Fe_3O_4)-based blood flow via tapered complex wavy curved channel with slip features. *Micromachines* **2022**, *13*, 1415. [CrossRef]
35. Waqas, H.; Oreijah, M.; Guedri, K.; Khan, S.U.; Yang, S.; Yasmin, S.; Khan, M.I.; Bafakeeh, O.T.; Tag-ElDin, E.S.M.; Galal, A.M. Gyrotactic motile microorganisms impact on pseudoplastic nanofluid flow over a moving Riga surface with exponential heat flux. *Crystals* **2022**, *12*, 1308. [CrossRef]
36. Shahid, M.; Javed, H.M.A.; Ahmad, M.I.; Qureshi, A.A.; Khan, M.I.; Alnuwaiser, M.A.; Ahmed, A.; Khan, M.A.; Tag-Eldin, E.; Shahid, A.; et al. A brief assessment on recent developments in efficient electrocatalytic Nitrogen reduction with 2D non-metallic nanomaterials. *Nanomaterials* **2022**, *12*, 3413. [CrossRef] [PubMed]
37. Mamatha, S.U.; Devi, R.L.V.R.; Ahammad, N.A.; Shah, N.A.; Rao, B.M.; Raju, C.S.K.; Khan, M.I.; Guedri, K. Multi-linear regression of triple diffusive convectively heated boundary layer flow with suction and injection: Lie group transformations. *Int. J. Mod. Phys. B* **2022**, *in press*. [CrossRef]
38. Kiranakumar, H.V.; Thejas, R.; Naveen, C.S.; Khan, M.I.; Prasanna, G.D.; Reddy, S.; Oreijah, M.; Guedri, K.; Bafakeeh, O.T.; Jameel, M. A review on electrical and gas-sensing properties of reduced graphene oxide-metal oxide nanocomposites. *Biomass Convers. Biorefinery* **2022**, *in press*. [CrossRef]

MDPI
St. Alban-Anlage 66
4052 Basel
Switzerland
www.mdpi.com

Bioengineering Editorial Office
E-mail: bioengineering@mdpi.com
www.mdpi.com/journal/bioengineering

www.ingramcontent.com/pod-product-compliance
Lightning Source LLC
Chambersburg PA
CBHW061559210926
51850CB00009B/2152